# Tawang, Monpas and Tibetan Buddhism in Transition

M. Mayilvaganan · Nasima Khatoon ·
Sourina Bej
Editors

# Tawang, Monpas and Tibetan Buddhism in Transition

## Life and Society along the India-China Borderland

Springer

*Editors*
M. Mayilvaganan
National Institute of Advanced Studies
Indian Institute of Science campus
Bengaluru, Karnataka, India

Nasima Khatoon
National Institute of Advanced Studies
Indian Institute of Science campus
Bengaluru, Karnataka, India

Sourina Bej
National Institute of Advanced Studies
Indian Institute of Science campus
Bengaluru, Karnataka, India

ISBN 978-981-15-4345-6 ISBN 978-981-15-4346-3 (eBook)
https://doi.org/10.1007/978-981-15-4346-3



This Springer imprint is published by the registered company Springer Nature Singapore Pte Ltd.
The registered company address is: 152 Beach Road, #21-01/04 Gateway East, Singapore 189721, Singapore

# Preface

I did not have any information about the people, society or environment in that region like other Indians, till our research study, as it was a remote, isolated and unknown place for many, in India. And I and my team did not envisage that we would ever visit that part of the world. The research study on Monpas and Tawang between 2017 and 2019, in fact, aided in bringing me and my research team to this wonderful borderland, a spiritual heaven lies 10,000 feet above sea level with an atmosphere of infinite peace, calm and beauty amid snow capped mountains coupled with the colourful monasteries and the smiling local Monpas with the low rumbling Buddhist chants. Tawang had always fascinated me since then.

Tawang, the name of the place, is believed to have been derived after the dazzling Tawang Gompa (Buddhist Monastery). In the local Monpa dialect, "Ta" means Horse and "Wang" means Chosen, i.e. sanctified horse chosen place. The local elucidation is that in the seventeenth century, a Tibetan monk called Mera Lama found a place, where Tawang Gompa stands today while he was looking for the horse. It is at the southern outer edge of Tibet, bordered by China on the North and Bhutan at the South-West and Assam plateau in the South. The Mongoloid featured Monpa tribes are the natives of Tawang, evidently closer to the people of Tibet in their origin. In Verrier Elwin words, Monpas are as a "gentle, friendly, courteous, industrious, good to animals, good to children". To get to Tawang, one has to take a jeep or car from Tezpur, a roughly two-day-long journey through the spectacular Se La (Pass), which is at a height of over 13,000 feet above sea level.

After the Tibetan administration, the area was under the British ruled North-East Frontier Tract (NEFT). Later, following India's independence in 1947 it became part of the North-East Frontier Agency in 1951 and with the formation of Arunachal Pradesh since 1972. However, this peripheral area became a strategic importance only subsequent to the Chinese military invasion in 1962 and its constant claim of Tawang as part of its southern Tibet territory thereafter in the last few decades. Consequently, the Indian government reinforced its military defences since 1962 and have undertaken borderland development.

With the research study on Monpas and Tawang, we convened a round table in the National Institute of Advanced Studies (NIAS) titled "Focusing the Frontiers: Understanding the Borderland Spaces, Identity, Perceptions and Imaginings of Monpas in Tawang in India-China Border", perhaps the first of its kind, exclusively on Tawang Monpas, in India on 18 December 2017. Around a dozen papers were presented by scholars expounding history, socioculture, environment, economy and geopolitics of the Tawang borderland and Monpas, as well as their ties with Tibet and Bhutan.

Along with some of the papers from the round table, our research study–Tawang borderland–brought excitement as a researcher to bring out this book's publication, as I felt it was extremely imperative to enlighten the public about this astonishing remote region. Not many books have ever been published about Tawang, and even fewer that have been published on the Monpa people and their area, mostly from the sociology or ethnobotanical or environmental study perspective. This is largely because the awareness about the area even to the Indian scholars is limited, and those who like to visit, even today, need to obtain special permission to enter the region. In a sense, this book presents a contemporary and multidisciplinary perspective on Tawang and Monpas. We hope that this book will add value to the existing literature on the subject and besides aid in introducing the area and the Monpa people to the Indian and the global audience in general.

Bengaluru, India M. Mayilvaganan

# Acknowledgements

This book entitled *Tawang, Monpas and Tibetan Buddhism in Transition—Life and Society along the India-China Borderland* is an element from a major research study on *Borderland and the Strategic Challenges: Towards Better Understanding of the Social Behaviour and Socio-Cultural, Religious and Economic Profile of Border People in Tawang*, at the International Strategic and Security Studies Programme (ISSSP) in National Institute of Advanced Studies (NIAS), sponsored by the Defence Research and Development Organisation (DRDO). So I would like to thank them. Notably, the book would have not been possible without the support of my research team, particularly Sourina Bej and Nasima Khatoon and all the contributors. I would like to record my sincere thanks to all the contributors for their cooperation at various levels, because without their support it would have been impossible to imagine the publication of this volume.

Besides them, there are numerous other people who supported me in this endeavour. I wish particularly to thank Prof. Rajaram Nagappa, the former Head of ISSSP, NIAS, for his insightful guidance and suggestions, and Mr. Yeshi Wangchu, Mr. Lham Tsering, Mr. Nima Dorjee and Mr. Tashi Tendar for the local support at Tawang, Bomdilla, Itanagar, and Bylakuppe. I would also like to extend my thanks to my colleagues in the ISSSP, Dr. V. S. Ramamurthy, former Director of NIAS and Dr. Shailesh Nayak, the Director of NIAS for their support and encouragement.

Furthermore, I would like to express my gratitude to the anonymous reviewers for his/her valuable comments and suggestions, which certainly aided in enhancing the draft. I also convey my earnest gratitude to Mrs. Satvinder Kaur, Ms. Sushmitha Shanmuga Sundaram and the other members of the Springer publication team for the timely assistance; without their cooperation, we would have not completed the book volume in time.

Last but not the least, I thank my family members (Dr. Amudha, Ms. Deekshitha, Mrs. Muthulakshmi, Mr. Ganapathy and Mrs. Jyothi) along with the family members of Sourina Bej (Mrs. Baishakhee Bej and Mr. Samir Bej) and Nasima Khatoon (Mrs. Nargis Begum and Mr. SK. Nizamuddin) for extending the moral support throughout the research study, writing and editing of this book.

Bengaluru, India M. Mayilvaganan

# Contents

# About the Editors

**M. Mayilvaganan** is Associate Professor in the International Strategic and Security Studies Programme at National Institute of Advanced Studies (NIAS), Bengaluru. He was a former visiting senior lecturer at the Department of International and Strategic Studies, University of Malaya, Malaysia, where he taught South Asia in International Relations, Modern Warfare and International Politics and formerly Associate Fellow at the Institute of Defence Studies and Analysis (IDSA), New Delhi. Mayil holds a PhD from the School of International Studies, Jawaharlal Nehru University (JNU) and has spent time as a researcher in Canada and the US. His research interests include strategic and security issues concerning India and its neighborhood, Indo-Pacific, foreign policies of India, China, Japan and the US, Borderland and non-traditional security issues.

**Nasima Khatoon** is a Research Associate in the International Strategic and Security Studies Programme at National Institute of Advanced Studies (NIAS), Bengaluru. Her areas of research include nuclear diplomacy, national security, and West Asia.

**Sourina Bej** is a Research Associate in the International Strategic and Security Studies Programme at National Institute of Advanced Studies (NIAS), Bengaluru. Her research interest is on water governance, climate change, identity and development politics in South Asia. She has also worked as a correspondent-cum-copy editor at The Times of India, Chennai.

# Chapter 1
# Life and Society on the Edge of Tibet

**M. Mayilvaganan**

## Tawang: The Land of Monpas

The Monpas are a predominant tribe in the districts of Tawang and West Kameng in the western highlands of Arunachal Pradesh bordering Tibet in the north, Bhutan in the west, and the plains of Assam in the south. They are said to be from Tibet and migrate to down south, i.e. Tawang, what was known earlier as Monyul in different periods of time. In fact, the term "Monpa" is apparently from the Tibetan language with, "Mon" meaning low or down the land, and "Pa" meaning people.

They were under the Tibetan authority and paid tribute to the rulers at Lhasa in Tibet till 1951 when India took control of administration in Tawang.[1] The high snow-capped mountain peaks, dense forest, rivers, stunning lakes, and waterfalls, combined with the extreme climatic condition made access to their area extremely difficult.

Besides, the oral tradition of the history of the Monpas is associated with certain myths and legends. Even today, the musings of these legends are revered by the Monpas; many revolve around Tsangyang Gyatso, the sixth Dalai Lama, who was born in Urgyeling, a village south of Tawang and the great tantric master and Yogi Guru Padmasambhava. Tawang is home to one of the most important monastery (gompa) of Tibetan Buddhism outside Tibet, which influence the spiritual life of Monpas. The Monpas regard the authority of the Dalai Lama and him as their supreme spiritual leader, owing to historical, cultural, and religious interaction that have crossed path since antiquity and contributed to building the collectives of the Monpas. Evidently, traditional Tibetan culture even now runs strong in Monpa land, Tawang.

At present, nearly the 49,000 Monpas, also known as the Menba or Monba, lives in Tawang,[2] apart from about 78% living in the adjacent district of West Kameng, particularly Dirang (Central) and Kalaktang (Southern) areas. They are the follower

M. Mayilvaganan (✉)
International Strategic and Security Studies Programme, National Institute of Advanced Studies, Bengaluru, India
e-mail: mayil@nias.res.in



M. Mayilvaganan et al. (eds.), *Tawang, Monpas and Tibetan Buddhism in Transition*,
https://doi.org/10.1007/978-981-15-4346-3_1

of the Gelugpa sect of Tibetan Buddhism compare to the other tribes in the state who practise obscure forms of animism. Also, Monpas have unique dialectical variations, customs and customary laws, dress and food habits.

The social structure of Monpa society is basically arranged or classified on the status of respectability, education and wealth, unlike on caste or class. A religious, knowledgeable, and educated Monpas such as *geshe*, monks, and *anni* occupy a high position in the society and respected. Also, the Monpas society is classified on the regional line. The mangma, traditional self-governing village councils of the Monpas and headed by *gaonburha* or *gaon burra* (the village head man) is in charge of the internal administration of the village that consists of about 20–200 households. The Monpa society follows several social norms and traditional rituals related to birth, marriage, and death of a person.

The Monpas live as joint families usually and even if it is nuclear family, they choose to live in the vicinity of other families. Marriages are highly preferred within the same group and the village level. The practise of monogamy is the common norm of Monpas marriage, however, polygamy too exists. The cases of brothers marrying a woman are evident among yak rearing high altitude nomadic Monpas (Brokpas). Also, male Monpas (sons) traditionally inherit an equal share of parental property, whereas women Monpas (daughters) are not entitled to inherit property. Customarily, the middle child (second son) in the Monpas family having three or more children should go to Tawang Gompa/monastery to be a monk.

The Monpas have a simple lifestyle, mostly grazing the yak, rear, and shear the sheep, collecting edible plants, and cultivating crops or involved in barter trade. Monpa women generally engage in carpet making, one of the indigenous crafts, weaving and agricultural related activities, whereas Monpa men are usually engaged in making of masks and images, and wood carving. Some Monpa men engaged in barter trade with their neighbouring Monpas in Tibet and Bhutan on the items like food grains, livestock, animal skin, medicinal herbs, wool products, and butter/ghee, etc. During festivals and leisurely hours, the Monpas enjoy playing games, sports, and dance dramas. *Losar*, observed in the first month of the lunar calendar, generally falls by the end of January or in the early February is the main festival of the Monpas.

In simple language, they are close to nature and mainly dependent on nature, their surroundings, for their day to day needs. Occupational mobility among Monpas is restricted to taking yak for grazing (mostly taken to high altitude areas and the possessor (Brokpas) stay there during the summer months.) and collecting food and woods. Mobility related to education was almost negligible in the past. The landholding was under the control of community or gompa (monastery). They have a higher degree of solidarity among the clan and respected traditions and customs, which binds them together. In general, their nature of shyness and lack of contact with the outside world, beyond Tibet, Bhutan, and Assam plains to an extent have aided in keeping their solidarity and customs intact.

## Life and Society in Transition

Nonetheless, today the life of Monpas and their society is gradually in transition with the fast-changing socioeconomic conditions in the region, something that has long been an ongoing process in the human society. But the pace of change has been very slow. Traditionally, the Monpas had adopted a self-reliant and subsistence economy with little ideas on means to generate a cash dependent economy. It is known that whatever they produce is for consumption and not intended for making profit. At the moment, due to the impact of tourist inflow, education, globalisation, and network development, Monpas are tending towards new economic order in the society. Every younger Monpa wants to earn quick money and well settled early with modern and luxurious lifestyle, thereby eroding the traditional way of resource dependency for tribe's persistence. Thus, the commodification of the nature has led to large scale unbridled infrastructural development and made way for easy returns of investment at the cost of maintaining sustainable ecology.

Today, many Monpa children are taking to modern education over monastery education with a belief that it is easier to obtain economic advancement through education and business than adopting traditional livelihood means. Modern education coupled with traditional monastic higher education at Bylakuppe in Karnataka certainly appears to be an instrument of social change within the Monpas younger generation.

The Monpa language, a vocal identity of the society in Tawang and the mother tongue of Monpas, however, today as the region opens to welcome tourists monolingual is slowly make way for bilingual use and given the pan identity of the usage of Hindi. Interestingly, it has emerged as the most preferred and comfortable lingua franca for the younger generation Monpa. Likewise, totally contrast to the local traditional culture, the majority of Monpas have completely switched over to the modern dresses, they prefer to wear their own traditional dresses only during the festival times. It appears that at least a decade ago many Monpas wear traditional dress often, but right now their dressing sense is as common as other people in India.

In terms of food habits, Monpas generally depend fully on nature from where they gather food and medicine. Their staple food is generally a simple one with mostly boiled and less oil dishes that has cheese, chilli, rice, maize, millet, buckwheat, meat, and vegetables. But evidently there is a gradual change today where the most of the younger generation are attracted to the cosmopolitan food culture like processed food, fast foods, burger, coffee, cappuccinos, apart from the mainland food items like puri bhaji (deep-fried bread made from whole wheat flour), roti (flatbread made from wholemeal flour), and dosa (a flat thin flayered bread made of fermented rice and lentil batter).

Another perceptible change that is evident today is in their traditional art and crafts. Monpas have a unique heritage of art and craft, including Thangka painting and handmade papers that are part of their socio-cultural and religious traditions. With the time, many have undergone transformation, only a few, say less than 5% of the rural Monpas practices their traditional art and craft, whereas the majority of the younger generation are not interested in nurturing their skill in this matter. In addition, mostly the people of elder generation and middle age are eager and concern about their traditional art and culture. On the other hand, the young generation wooed by the western culture does not have the patience to learn their traditional handicraft. Understandably, the lack of market and the expensive nature of their traditional products too deter them as in most cases they become obliged to sell low price items to the tourist, who are the only customer to visit their region.

Also, Monpas who worship numerous local gods, deities, and spirits apart from the lord Buddha, where the religion is basically centred in this respect, nowadays with the spread of modern education and under the growing influence of urbanisation the younger mind's belief system is gradually fading away. Since most of them are exposed to modern society, ideas, and rational of questioning, the existing primitive beliefs of the clan have made them less interested towards traditional believe system.

The traditional landholding pattern of the Monpas too is gradually changing. The percentage of individually owned private land increases compares to community owned, with the rapid development associated with tourism and agriculture and horticulture. Also, currently, many hotels and guest house with modern facilities are built in Tawang and even some of the houses are renovated using the imported materials.

With the passage of time, the marriage patterns too have undergone some changes, and nowadays, marriage outside the community is taking place widely. Many of the younger Monpas prefer a nuclear family than joint families.

The changing of attitudes is usually expected with the change of generation. Monpas are no exception. On the one hand, they are trying to preserve their old tribal ideals of community, and on the other hand, they are forced to conduct their life in accordance with the socioeconomic change, market economy, and the concept of competition. Several factors have brought about changes in this tribal culture, such as measures undertaken by the government of India after the 1962 war, stationing of large number of troops, establishing communication facilities and education system, rapid infrastructure development, and finally the tourism.

The nature of change is such that Monpas are not losing their identity and traditional cultural heritage, but began adapting to few cultural traits of modern society. The younger generation Monpas who are exposed to modern education, telecommunication, and internet, etc. have developed an individualistic outlook by adopting new attitudes and behaviour patterns.

## Alienated Borderland to Fortified Borderland

The land of Monpas, Tawang borderland, is a space where life and society of Monpas were historically shaped by and the transnational and transculturation interaction is well-guarded at the moment. Once an integral part of Tibet administration, today Tawang is a well-fortified Indian borderland were routine cross-boundary interchange is practically non-existent owing to India-China political disputes and militarisation of the region. With China's claim of sovereignty over all of historical Tibet, Tawang too is being claimed as part of southern Tibet. Also, as Chinese grown increasingly hostile to the Dalai Lama, the exiled spiritual leader of the Tibetans, who passed through this borderland when he fled into exile to India in 1959, they protest his visit, if and when, to Tawang. And in some ways, Tawang has emerged as the focus of India–China relations or border dispute,[3] which revolves around the fulcrum of Tibet.

Keeping geostrategic considerations in mind, India accelerated border protection immediately after China's invasion of Tawang in 1962 and stationed large troops to guard. Strategically, Tawang is the most important division is the eastern Himalayan mountain chain since then. Military bases are in every half-mile in the border and considerable amount in the Tawang town. Incidentally, the disputes over borders and territory between India and China altered the lifestyle of peace-loving Monpas to an extent.

## The Core of the Book—The Subject Focus

The Tawang borderland in the Indo-China frontier is as dynamic as it is evolving in nature. But most of the understanding of this borderland gets trapped in the strategic prism that the larger boundary dispute between the two emerging powers in Asia (also shares a long boundary of over 3000 km) dictates. This edited volume tries to look at this strategic borderland not from the eyes of the dual power centres but keeps the borderland in the centre to understand the everyday life of the border community and the development that takes place in the geographically remote location.

This edited volume provides historical analysis on the borderland from being under the Tibetan administration to the events leading to the 1962 Sino-Indian war and to in the present day. The main objective of the book is to examine the nature and the current status of the life and society of the Monpas, the border community in the eastern Himalaya, on the edge of Tibet.

In particular, in this volume, the varied chapters have chosen to explore the history and socio-culture of Tawang borderland and the border people Monpas that sits at the cross junction of China, Bhutan and India. It served as the transit point for the seat of powers: the Bhutan kingdom to the west and the Tibet to the north (before the People's Republic of China exercised its administrative control over Tibet in 1951). Ever since the formal boundary line in the form of Line of Actual Control (LAC) was reinstated between India and China, this cross junction has ceased to be a transit and have been regimented into modern barbed wires. But beyond the rigid border, the reminiscence of the Tibetan and the Bhutanese influence is a part of the collective memory of the border community in Tawang even at present.

The Monpas have traditionally lived in the south of Lhasa, with a tradition of monastic education, environmental subsistence, and dairy dependent cuisines. Yet with their unique traditions and follower of the Gelug sect of the Tibetan Buddhism, they form a closer bond and draw its lineage to Lhasa. The tales of the exiled Dalai

Lama crossing over to Changlang and entering India in 1959 still strikes a chord among the Monpas in the Zemithang, Tawang and Bomdila valley in Arunachal Pradesh. This volume tries to take a critical look and documents the historical narrative that exists between Tawang and Tibet.

At the same time, coexisting with the history, the traditional subsistence way of life of the Monpas is undergoing a change brought about by the modernisation forces and the market economy primarily. The only accessible and option of monastic education at the revered seat of Gelug Buddhism, i.e. Tawang Monastery is today replaced by modern education. More and more youths are opting out of the once prestigious monk philosophy for reasoning and capital growth driven profession. The state government's policies to capitalise on the picturesque hills of Tawang and turn it into a tourist destination have also attracted the young Monpas to build and drive their entrepreneurial instincts. This has come at the cost of a little adjustment in their food, health, and environment. The volume explores this transition that has taken place in the close-knit societal, environmental, and religious and the economy of the Monpas. Simultaneously situated at the distant border, the lives of the Monpas are tied to the terms and conditions of the development that the central administration chooses to bring about thereby limiting their independent nature of subsistence. The state-driven infrastructural development brings with it not only technology and telecommunication advancement but when the community also accepts and adapts to the changes at the cost of the large swathes of the community-owned forested mountains that are brought under construction. This volume tries to examine this nature of development in Tawang borderland and the implication that it has on the Monpa's way of life too.

Finally, the volume also throws light on how the border community maintains its relation with the security agencies through border management.

In a sense, the edition brings out the contemporary issues facing the Monpa communities of Tawang, thereby furthering the knowledge of the dynamics and interactions between the state institutions, government policies, and the community. Particularly, this volume addresses the lacuna in understanding the gap in the history behind the nature and formation of the Mon identity and what sustains this identity till date? The book builds the new body of scholarship on the Monpas looking at the series of questions on the politics of belonging, their understanding on border and security, culture, environment preservation and its relation with Tibet.

## What are the Chapters all about?

The volume could be broadly divided into three sections. The first five chapters look at the historical narration on the formation of the Mon identity, the relation of Tawang with Tibet and lastly on, how did Tawang join the Indian administration post-1947 Indian independence. This historical build up necessitates the understanding on the nature of the borderland before the establishment of the modern state. It places the border district of Tawang in the context of why it is an important strategic borderland today and how has the Monpas historically responded to this narrative.

The next few chapters look at the internal dynamics in the Tawang borderland from development, infrastructure, and environmental way of life of the Monpas of the region. Situated at the border, the internal dynamics of Tawang cannot be completely isolated from the external dynamics that the region has come to play in the India–China border contestation. Thus, the last section of the book looks at this external dynamics and how the contention between India and China has evolved over the state of Arunachal Pradesh and where could Tawang be located in India's border management efforts.

In Particular, questioning the singular interpretation of the border, the second chapter by Sherin Ajin attempts to understand the India–China conflict as an evolving process that has made the border and the borderland equally dynamic in nature. The author locates the Tawang borderland in the domain of international relations theory where borders have come to define the modern state building and the sovereign sanctity which are preserved and protected. The author makes effort to understand the dynamic evolution of the border by the interactions within the borderland weaving in the historical intricacies. Author explores where "Tawang" does as a borderland situates itself in the larger border discourse in this process of Sino-India border dispute. The chapter goes on to describe that each border has a context of its own, and in order to understand the borderland, one needs to understand the warp and weft of that specific border. Throwing a caution towards building of a single narrative, the author brings out the risks of over simplification and reductionism in the diverse nature of the border.

Neeru Nanda's experience as the former additional deputy commissioner of Tawang comes live in the third chapter, wherein the author tries to see Tawang borderland from the Mon perspective. It is a historical overview of how the borderland developed that has significant linguistic and cultural affinities which are found among other Mon people from Ladakh to Sikkim, Eastern Bhutan, Tawang, and West Kameng. The answer lies in the way the Mons of Tawang chose their administrative umbrella after the sectarian rivalry between Gelukpa and the Drukpa sects of Tibet, the two sects of Buddhism that divided. While on the one hand, Bhutan was unified under the Drukpa rule, Tawang chose to stay with the Dalai Lama. However, later the extensive taxes by the Tibetans led the Monpas in Tawang and Kameng to extend warm reception to British administration. Post this, once again the Monpas chose to evade the Chinese control Tibet and joined India under Major Bob Khating effort.

Since then the Mon identity has taken a centre stage in deciding their culture, religion, and linguistic choices. And, all these have happened remaining under the strong regimented border and coexisting with the Indian Army. Interestingly, the author brings out a further trend—a new vision of cultural revival with the local cultural organisations set up by educated Monpas who started documenting the bon heritage of Tawang which ante-date the establishment of the Tawang Monastery (gompa). In an attempt to recreate the present through the past, the Monpas in Tawang are forging inter-linkages among the Himalayan communities from Ladakh to Lepchas of Kalimpong. Situating in the current political context, the author looks at why is it that the Monpa so confidently say, "We are neither the people of Gyaser (Tibet) nor the people of Gyakhar (India). We are the Mon". Being unique within the Indian polity, in fact, help them adjust comfortably to the so-called mainstream. The understanding on the Mon identity will remain limited if the relation between Tibet and Monyul is not drawn.

This story of legitimacy is further brought out by Claude Arpi in the fourth chapter titled "Standing at the Himalayan Crossroads: A Tale of the Liberation of Tawang". Drawing from Vallabhbhai Patel's letter to Jawaharlal Nehru, the author notes that the story of Tawang coming under the Indian administration goes to the foresight of Patel, who was then the Union Home Minister and the Major Bob Khathing who was given the task of leading the mission. Patel had listed some of the problems including Tawang in the post-independent India, which, according to him, required "early solution and around which we have to build our administrative or military policies".

In the next section, the chapters look at the internal dynamics that have come to dominate and impact the lives of the Monpas in Tawang today. The Tawang Monastery plays a big role in shaping the religious identity of the Monpas. Tsetan Namgyal in the fifth chapter titled, "Monpas, Tawang Monastery and Tibetan Buddhism: Ethno-Religious Links" attempts to examine and assess the ethnic identity and culture of Mon' and the "*Monyul*" and also the most noted monastic establishments of Tawang, while correlating the religious and political significance in the context of present geopolitical scenario. The historical significance of the Tawang monastery is that it forms a crucial part of Tibet's national identity in general and more than that it represents the local identity culture of "*Monpa*" ethnic people who constructed and developed it into a world Buddhist heritage.

The Tawang monastery that is situated in the historical region of "*Mon*" kingdom, which was incorporated into Tibet during the late seventeenth century by the Lhasa government, and today that is an integral part of India's state of Arunachal Pradesh in the eastern India's Himalayas. Reiterating what Neeru Nanda writes in her chapter, the author says all available Tibetan sources referred the region and the locals as a distinct ethnic group once with a separate region called "*Monyul*" and considered to be having some historical cross-cultural relations with Tibet, Bhutan, and Ladakh mostly through forceful migration.

The author brings out that Mon is not a monolithic ethnic identity. "Mon" is also one of the prominent and distinct ethnic groups of Ladakh and Baltistan region of northwestern Himalayan belt. What makes the Monpas of the Tawang different is how today, the traditional religious culture of the Mon and the Monpas is accentuated by the Tawang monastery, which is one of the largest Buddhist monasteries in the entire trans-Himalayan belt. While the Buddhist epitome of identity symbolises the Tibetan identity in exile, the monastery becomes a part of that larger identity. At the same time, the Tawang monastery is seen by China as the strong reason for its claim on Arunachal Pradesh (as Southern Tibet) because of its historical connection and linkages with three largest monasteries of Lhasa (*Sera, Drepung, and Gaden*) in particular and considering that all the three monasteries is a part of Tibet which now is a part of China. The chapter brings out the history of cross exchanges of identities that has existed in the borderland, shaped the religious identity of the Monpas as well as the counterclaims that each state and the communities have made time and again to own their social and political relations.

The unique culture of the Monpas is also reflected through their lifestyle and food habits, in the sixth chapter titled, "Lifestyle and Health Issues among the Monpas of Tawang", M Mayilvaganan brings out the transition taking place in the Monpa lifestyles and food culture and how the younger generations of Monpas are particularly the one at risk with the sedentary lifestyle and food habit changes today in Tawang. The chapter examines further the sociocultural settings in which the transitions are occurring in the Mon land and its influence on food and drink habits. Tied to the changing food habits, are also health challenges that the Monpas face in the present day where health infrastructure is at nascent stage.

The author highlights that it is important to note that the changes in lifestyle, food and health challenges are not in isolation to the socioeconomic changes. As more and more youths migrate outside for higher education, a cosmopolitan and business culture has come to mould the mundane lives of the otherwise close-knitted tribal identity of the Monpas. A shift towards power and money has in turn led to the commodification and certain new additions to the food of the Monpas. The changes are not only noted in the economic conditions but in the health sector too. Earlier only traditional healthcare was predominant but today the modern allopathic remedies are sought by many Monpas with the development of district health hospital and government health institutions.

While some changes are good, some has come at the cost of the environment and the subsistence nature of livelihood of the Monpas. Nasima Khatoon in the chapter seven, "Natural Resources and Biodiversity Conservation Practices in Tawang", points out that traditionally the Monpas of Tawang have adopted a natural and sustainable lifestyle with the reverence for environment and the surrounding biodiversity. But, with modernisation, this harmonious balance is facing a slow transformation. With a uniform market, government's growth-oriented economic policy and the large-scale infrastructural development occurring simultaneously in the district, the traditional economic practices are altered, and today from the building of hotels, roads, bridges, and dam, the environmental depletion and the harmfulness to the ecology is evident.

The chapter also focuses how these development push has slowly come to change the Monpa's own relation with the biodiversity where the community would not only depend on the forest lands for subsistence but also assess the threats to the ecology and design a conservation practice which is sustainable in nature. The chapter also analyse the rationale behind the push for infrastructure in the borderlands. While it does not take place only to fulfil the development deficit of the region, it is also determined by the historical significance Tawang holds in India's external policy towards China.

Mirza Zulfiqur Rahman in the chapter eight, "Infrastructuring Arunachal Pradesh Borderlands: A Case of Tawang Borderland", explores the several infrastructure projects undertaken by New Delhi in Arunachal Pradesh that has picked up pace in the past decade. The author points out that this is often seen as a concerted effort to strengthen the mobility of the armed forces up to border given that China contests India's claim over the territory of Arunachal Pradesh. These development projects are often politically portrayed within Arunachal Pradesh as priorities to address the huge developmental gap that the various tribes in Arunachal Pradesh live with.

In addition, the chapter examines the dynamics of these infrastructural projects in Arunachal Pradesh, and its utility in hardening of the borders, disrupt traditional people-to-people exchange routes and sacred sites/spaces in borders and employing hydropower dams as an instrument of legitimacy over Arunachal Pradesh. It is interesting to note that in 1962, the legitimacy depended on the Monpas' exclusive decision to accept the nature of control, but in contemporary times, this exclusivity and the autonomy to choose the type of development has eroded.

But how beneficial are the hydropower projects that are being developed in Tawang? and its impact on environment. In the ninth chapter by Amit Mukherjee, "Hydropower Projects in Tawang: Concerns and Flood Proximity Estimation Using GIS Analysis", the author attempts to address the major concern of flooding that is deemed to be caused by the building of dam projects on the rivers of Nyamjang Chu and Tawang Chu. In the course of the analysis, the author first looks at the proximity to the inhabited places followed by the altitude as to whether in natural conditions, it can raise the water levels up to these villages/towns. As such visual interpretation and inference have been carried out using standard geospatial methods of digitization. Cross-sectional maps were then generated to identify and analyse relevant topography and its suitability for building of dams to carry out the analysis.

Even before the completion of the projects and any major study that proves the detrimental nature of the dams based on scientific ground, the anti-dam protests against the two hydropower projects on the Nyamjang Chu and the Tawang Chu rivers have begun in Tawang in early 2010s. In the chapter ten, Sourina Bej looks at the local protests and their differing dynamics. The chapter chooses to understand how a local protest that is born out of the grass root systemic mismanagement of resources slowly transforms to acquire a national or transnational dynamic owing to the location of the dam on a geostrategic border. The protest by the Monpas is not different from the discourse on the politics of dam in India but it unique in its narrative, causes, and the tools used to legitimize their voice.

The author looks at the rhetoric that the anti-dam protests by the Monpa make and how different is it from similar local resistances in Dibang, Lower Subanshiri, Ranganadi, and Loktak in the Northeast India. Fears of migration, environmental depletion, and the loss of land are the primary reasons dictating the internal causes of the protest. The social movement began against the dam projects has now opened up the democratic space for the Monpas to voice their dissent. The author points out that counter narrative also plays an important role in understanding how the Monpas and the state claim who they are, their land relations, and their resource relations. The local protests in Tawang are on two rivers that have differed depending on the community's dependence on resources and their relation with the land and the forest. While at the core, the internal issues of socioeconomic and religion remain the reason for mobilising the resistance against the dam, the historical memory of the 1962 war in Tawang gives an external interpretation to the protests.

Just like the politics in Tawang cannot be looked in isolation from its external geographic location, similarly the geostrategic importance of Tawang is intrinsically linked with how China has come to locate Arunachal Pradesh in its larger territorial debate with India. In the eleventh chapter, Jabin T Jacob looks at Arunachal Pradesh in the India–China relations focusing on the trends in Chinese behaviour and their implications. The centrality of Arunachal Pradesh in the relation between the two countries has increased in profile over the last few decades with key issues being transgressions by Chinese troops across the Line of Actual Control, the changing Chinese practice in terms of issuing of visas to Arunachali citizens, and the visits of the Tibetan leader Dalai Lama to Tawang.

The author examines each of these issues briefly before focusing on a 2017 Chinese exercise of renaming places identified as falling within Arunachal Pradesh. India has responded in its own way to deal with the issues but is often limited by the structural deficits or the stagnated institutional decision-making process. The author explores the evolving nature of the relation and how the importance of Arunachal Pradesh in relation might shift in the long run. While for China Arunachal Pradesh may today be at the heart of the Sino-Indian boundary dispute, but the dispute itself is increasingly less at the heart of the Sino-Indian relationship. And this appears to be the case also for India.

The security and management of the border are increasingly dependent on a shared understanding and relation with the local border community, the final chapter by Pushpita Das, titled, "Border community in Border management: A case of Tawang Tract", examines the role of the border community in the border management. The author states that Tawang has remained in the periphery first as the South Monyul and now at the periphery of Arunachal Pradesh in the larger boundary dispute with India and China. Hence, the local community has balanced the dual power centres, first with Tibet and Bhutan and now with India and China. Their existence is based on how they managed their relation with Tibet and later by maintaining their exclusivity with India to evade the harsh taxations by the Tibetan official and lastly by giving passage to Dalai Lama when he escaped Tibet. The Indian government has also tried to instil a sense of loyalty among them through various training programmes. But in recent years, a sense of neglect is creeping among the local people given the pervasive underdevelopment in their area. The version of infrastructural development pursued by the state has not overcome the sense of neglect, rather leading the people to demand separate autonomous district council to preserve their identity.

Further, the author identifies the significant role played by the Monpas in communicating and trusting the stationed troops and giving timely information to the border on any sort of incursion. Indeed, Monpas has been extremely wary of China because they have heard the narratives on Chinese atrocities on Tibetans across the border. The author concludes by stating three ways in which relationship between local and the security personnel could be developed: first, sustained community interactions by establishing committees comprising the security personnel, village headman and prominent persons; second, local communities should be trained to act as eyes and ears for the security and intelligence agencies deployed at the border; and last, the local people should be empowered and their participation should be elicited in the planning and execution of various development schemes, so that they become active stakeholders in a peaceful and secure border.

## Endnotes

1. After the treaty of Yandabo in 1826, the British Government took over the administration of Assam and the North-Eastern Frontier Tract. And later with 1914 treaty (Simla Convention) with Tibet and McMahon Line, Tawang came under the British-run Indian jurisdiction. Shortly, after independence, at the instance of the Government of India, Major R. K. Khating also called Bob, the Naga officer of the Indian Frontier Service and then the Assistant Political Officer at Charduar, who trekked to Tawang on the order from the government, hoisted Indian flag in 1951.
2. Government of Arunachal Pradesh, District Demography, as per the provisional figure of 2011 census, at https://tawang.nic.in/about-district/demography/.
3. K J M Varma, "Concession on Tawang can resolve India-China border dispute: Dai Bingguo," *Live Mint*, 03 March 2017 at https://www.livemint.com/Politics/bTPTWh5efcWfdYGW6VMNLL/Concession-on-Tawang-can-resolve-IndiaChina-border-dispute.html.

**M. Mayilvaganan** is Associate Professor in the International Strategic and Security Studies Programme at National Institute of Advanced Studies (NIAS), Bengaluru. He was a former visiting senior lecturer at Department of International and Strategic Studies, University of Malaya, Malaysia, where he taught South Asia in International Relations, Modern Warfare and International Politics and formerly Associate Fellow at the Institute of Defence Studies and Analysis (IDSA), New Delhi. Mayil holds a PhD from the School of International Studies, Jawaharlal Nehru University (JNU) and has spent time as a researcher in Canada and the US. His research interests include strategic and security issues concerning India and its neighborhood, Indo-Pacific, foreign policies of India, China, Japan and the US, Borderland and non-traditional security issues.

# Chapter 2
# Situating the Sino-Indian Border of Tawang in the Border Studies Discourse

**Sherin Ajin**

## Introduction

Imageries of border, in the realms both ideational and material, appear disruptive and fracturing the supposed seamlessness of the whole. Inhabiting a fragmented globe, where one's location within the specificities of latitudes and longitudes determine one's spatial, temporal and ancillary associations, one dwells in the mutiny of fissured imaginations. To put it simply, we "inhabit many borders".[1]

The geography of border is embedded in the ambiguity of in-betweenness. Interestingly, this trait of ambiguity is reflected in the appellations used interchangeably for border like frontier, boundary, borderland, hinterland, etc. A simple exercise of scanning dictionary for understanding the distinction between these terms would reveal a mutual inclusivity of definitions.

Both the theory and praxis of border are deeply entrenched in the logic of binary, which in turn creates the dual worlds of "us" and "them", "insider" versus "outsider", etc. More than often, the process of border making is permeated by circuits of power. To explain it further, the geometry of border simultaneously performs the contradictory functions of exclusion and inclusion, exemplified in the Simmelian metaphoric "door".

Popularly understood as a way of "ordering space", the technologies of division and its tangible markers such as fences, barbed wires and walls perform the obvious functions of marking, limiting and as a logical corollary defending the thus produced bounded space. Though borders are not truly impermeable, transgressions or deviations are meted out with punitive mechanisms such as deportation, incarceration, torture and sometimes even death. Therefore, with its technologies of division and governance, border becomes an active site of biopolitics.

S. Ajin (✉)
National Institute of Advanced Studies, Manipal Academy of Higher Education (MAHE), Bengaluru, India
e-mail: sherinelizabeth03@gmail.com



M. Mayilvaganan et al. (eds.), *Tawang, Monpas and Tibetan Buddhism in Transition*,
https://doi.org/10.1007/978-981-15-4346-3_2

Also, interestingly, unlike many other conceptual engagements, it is difficult to agree upon a single, all-encompassing theory of border. As Ansi Paasi argues "*a general border theory seems unattainable, and even undesirable, individual state borders are historically contingent and characterised by contextual features and power relations. There can hardly be one grand theory that would be valid for all borders*".[2]

This chapter would be anchored on the understanding that each border has a context of its own and therefore any attempt of making sense of it needs a sensitive interpretation of the innate warp and weft of that specific border. This is because any efforts towards creating a master narrative bear the "risk of reductionism", thus, obfuscating the nuances weaved into the tapestry of every border.

Further, semantically and otherwise, the very term "border" is embedded in statist imagination, where the idea and manifestation of border becomes a spatial strategy through which people and resources are to be controlled to meet the statist ends. In other words, it intends to create and impose a certain kind of imagination about self and the world around. Yet the experiential realm of being cannot be limited to the prism of border alone. What is identified as the space of border is also the space of everyday living, which has a distinct socio-cultural and even ecological bearing. Beyond the statist territorial imaginations, there exists a realm of meaning-making and existence. For instance, in the case of Tawang, apart from its strategic geographical location, it also stands at the confluence of a distinct cultural understanding of Himalayas which is intertwined with its faith system. For instance, to the Tawang Monpas, what state imagines to be the riparian resources pertains to the realm of sacred.

A placid district in the Indian state of Arunachal Pradesh, sharing borders with China and Bhutan, Tawang sits at the locus of an intransigent border dispute between India and China. Second only to the Potala Palace of Tibet, Galden Namgyal Lhatse of Tawang is an esteemed seat of reverence in the Tibetan Buddhist circuit. Historically, Tawang has witnessed change of power flags from theocracy to democracy, a transition that has not garnered much academic pursuit. Also, Tawang is a classic case of a space that gets subsumed in the trappings of the territorial imagination of border, with its parallel processes of everyday living and meaning-making pushed into the backseat.

## Tracing the Border Discourse: An Interdisciplinary Narrative

In the pre-state temporal context, the frequent conquests and loss of territory closely resembled a political soccer. Devoid of modern-day concepts like sovereignty, the period is marked with shifting boundaries and therefore the territorial corpus mimicking an amoebic geometry. With the constant fluctuation of the contours of the limits of political power, the pre-state borders remained largely fluid, defying certitude.

A brief survey of the evolution of border studies starts with Friedrich Ratzel's 1897 classic titled *Politische Geographie,* where Ratzel identified the state with the merits of a living organism, ingrained with the potential of expanding and enriching itself. Hence, for Ratzel, boundaries of the state were not to possess the characteristic of *stasis.* In his work, "*he put forth an exceptional myth about the "organic" relationships between* ***volk*** *(people),* ***boden*** *(territory) and* ***staat*** *(state) and introduced the notorious concept of* ***lebensraum*** *(living space)*".[3]

Inspired by this conception of state and border, Rudolf Kjellen furthered the Ratzelian model through his work "*The State as a Living Form*". To Kjellen, the geopolitics translated to "*the theory of the state as a geographical organism or phenomenon*".[4] Premised on the Ratzelian body of work, the first functional distinction of borders as "good" and "bad" came from Otto Maull. While good borders were in synergy with other markers of boundary such as nature, culture, bad borders were non-aligned with the existing, non-political boundaries. It was Whittermore S. Boggs who gave the Maullian binaries an utilitarian twist based on the broad principles of "maximum efficiency and minimum friction".[5]

Meanwhile, Halford Mackinder's heartland theory and Alfred Thayer Mahan's theory of sea power garnered attention and invited a mutiny of debates. With his contested Rimland theory, Nicholas J Spykman encountered Mackinder's fetishism with "inner crescent".[6] To Spykman, it was the periphery that was of significance than the inner "core".[7] On the other hand, in the French intellectual circuit, Ratzel's organic, deterministic model did not gain much ground. Political geographers like Paul Vidal de la Blache tried to understand the relationship between man and nature.[8] Blache explored the potential of human agency in choosing socio-spatial possibilities. Pushing the possibilist thought, Lucian Lefebvre portrayed man as the "*master of possibilities*",[9] capable of manoeuvring the nature given possibilities to his use. Breaking away from the Vidalian understanding, it was Elisee Reclus who first introduced the term *geographie sociale.*[10] Perceiving space as a social product, Reclus emphasised the possibility of a dynamic understanding of the concept of space, with processes of meaning formulation, contestation and recreation embedded in the everydayness of life.

With his nodal work titled, *La Production de L'Espace,* Henri Lefebvre set the dialogue in motion understanding the "social production of space".[11] To Lefebvre, space is inevitably entwined with the processes of the social realm, tremendously influencing the configuration and practices of space. The French Marxist Sociologist famously made the tripartite distinction of space into (i) perceived space, (ii) conceived space and (iii) lived space.

Inspired by the Lefebvreian model, Brenner and Elden mapped the spatial category of territory upon the above-mentioned triangular distinction of space. And this superimposition meant division of the space of territory into the realms of: (i) territorial practices, (ii) representations of territory and (iii) territories of representation. To Brenner and Elden, the space of territory, including the borderland, is an active theatre of lived experiences, evolving itself in interaction with the social processes of its environment and even beyond it.

In the meantime, there were efforts in incorporating scientific rigour in the discipline, resulting in the birth of what could be called "border scientism". The works of Walter Christaller, August Loch and Torsten Hagerstrand discounted the Durkheimian sociality and instead focused on the objective aspects of spatial patterns, border scientism and underscored the limits of constructivism and human agency in landscaping the contours of the space of borderland.

Even though, the trajectory of border studies largely remained ancillary of few disciplines like political geography, history, etc. it was equally influenced by the dominant schools of thought of its times such as positivist, behaviouralism, functionalism, etc. Scholars like Richard Hartshone tried locating the space of border in the functionalist framework at the point of contact between the political and cultural scapes. As a corollary, he attempted what could be called the "genetic classification of borders" (Laine 2015), distinguishing "*borders as pioneer, antecedent, subsequent, consequent, superimposed or relic*".[12]

Hartshone was succeeded by Ladis Kristof, who in his article entitled "*The Nature of Frontiers and Boundaries*", tried to distinguish between "*frontiers*" and "*boundaries*". To him, "*etymologically, frontiers*" refers "*to that which is in front*", "*that part that was ahead of the hinterland*", with an "outer orientation", displaying "centrifugal tendencies".[13] In other words, frontier was not to be associated with the geographical periphery, as often popularly interpreted. On the other hand, according to Kristof, "boundaries marking the political and geographical limits of the state were "*inner oriented*", performing "*centripetal functions*".[14]

To functionalists like Minghi, it was crucial to broaden the horizons of borders, understanding them beyond the symbolisms of demarcation, delimitation and political divisions. Boundaries were to be approached as complex spaces of multifaceted interactions like social, cultural, economic, etc.

This being the concise, not conclusive, course of evolution of the discipline till the 1970s, the onset of cold war made the largely apolitical and objectivity obsessed canvas of the discipline look parochial. Echoing the intellectual disruptions of the time such as post-modernist and constructivist schools of thought, what existed as a dependent area of academic exploration began carving a niche for itself. The process marked a significant shift in its focus from "boundaries" to "borders". Thus, now, border was not to be singularly perceived as the geometry of territorial boundedness, but as an eclectic and exciting space that socially produces, defines and redefines itself.

Further, the discipline of border studies has meandered through a diversity of ideational and empirical landscapes shaping a variety of approaches. These approaches could be classified as traditional, typological and functional apart from post-modern, geopolitical, social representation policy-perception-practise (PPP) and eco-political approaches.

Traditional approaches to a large extent confined themselves to the larger morphological and immediate empirical analysis of border. Some of the approaches that fall within the signpost of traditional are: "historic-cartographic, typological, functional and the geopolitical approaches".[15] The major preoccupation of the historic-cartographic approach has been about the demarcation and delimitation of the boundaries. Typological approach laid emphasis in documenting and classifying the borders according to their tangible traits. Devising parameters of distinctions, different types of borders were identified, such as "antecedent, subsequent, colonial, post-colonial, phantom and borders with varying degrees of legitimacy".[16]

An offshoot of the post-World War II milieu, functional approach explored the nature of interaction between the socio-politico-territorial elements of both sides of the border. One of the interesting contributions of this approach being the identification of the simultaneous and diverse functions of the border as a "barrier, contact and filter" (Havlicek et al. 2018). In the realm of realpolitik, border controls the degree of contact, inclusion and exclusion, primarily based on its definitions of threat and the desirable corpus of the self. Premised on the nature and degree of contact, Oscar Martinez distinguished borders into "alienated, interdependent, integrated and frontal borders".[17]

Anchored in the disciplinary domain of political science, the larger agenda of the political approach was to map and locate borders in their functional sense as the sovereign limits of the state as in the prominent schools of thought such as realism, liberalism and globalism. Depending on the choice of framework, particularly the positioning and understanding of the idea and practice of state, each school engages with border differently. On one end of the pendulum, in the realist imagination, border becomes the sacrosanct space, defining the state that cannot be compromised. On the other end, in a highly interconnected world of networks, in the globalist imagination, the space of border is no longer impermeable and the regimented state systems giving way to an increasingly borderless world.

Inspired by dependency theorists such as Paul Baran and Andre Gunter Frank, Immanuel Wallerstein's world system theory[18] employed the categories of core, semi-periphery and periphery in understanding the unequal spatial distinctions permeated by the hierarchies of power and therefore development. Charting the corpus of state, in an obvious geographic sense, border constitutes the periphery. On the otherhand, power asymmetries created by the core/centre culminates in unequal resource allocation, etching patterns of regional inequality. Ironically, hence many a times, geographical periphery becomes a periphery in substantive sense as well.

Similarly, Anthony Gidden's structuration theory[19] which gives primacy to the interactions between the structure and agency without any disposition to pre-conceived hierarchies has inspired border studies scholars in employing the structuration framework in analysing borders of their specific interest. 9/11 attack, coupled with the emergence of non-traditional security challenges and their gravitas rejigged the global and national threat perceptions and agendas, initiating a plethora of literature revisiting the idea and practise of border against the new, abstruse security conundrum.

Another approach that has gained momentum in the disciplinary field of border studies is the policy-perception-practise (PPP) approach. Essentially, this approach attempts to merge two ways of seeing the border, that is, (a) as an "*international, legal institution*"[20] and (b) a product continuously shaped by the socio-politico-cultural perceptions and practises of the people who inhabit the border. Aiming to bridge the deficit arising from the isolated understanding of the concept of border, PPP tries to understand the experiential dimension of borderland population in oceanic/concentric circles, identifying the barrier function of the border to be disrupting the markers of ascriptive identities, such as ethnic and cultural.

Beyond the immediate political and strategic significance of the spatiality of border, ecology constitutes an integral part of its identity and existence. At the present temporal juncture, non-traditional security challenges such as climate change and pandemics are not limited to any specific geographical boundaries and therefore effective redressal demands sensitive and concerted efforts. Political ecology approach of studying the borders puts premium on this understanding of the spatial constitution of the border, attempting to address some of the pressing concerns of the times.

Incidentally, the idea and practise of border as is understood today has its taproots in the modern Westphalian state. Though there are debates on the historical accuracy of attributing birth of modern state to the Treaty of Westphalia, the 1648 Peace of Westphalia is widely considered as a nodal milestone, ushering in the foundational premises of modern state and thereby the idea of border. The Treaty/Peace of Westphalia which drew curtain on the gruesome annals of European history prescribed an alternative political organisational possibility of "international system", firmly anchored in the concept of sovereignty. This act of charting a way out of the perils of an anarchic world was premised on the logic of a less ambiguous delineation of sovereign, territorial states. In other words, "territory" became a centrepiece of the political spatiality of this schema.

Hence, as a logical corollary, in the Westphalian framework, the geodesy of border becomes the marker of the limits of the sovereign power. Taming and disciplining mobility, state borders act as "*territorial markers of the limits of sovereign political authority and jurisdiction, located at the geographical outer edge of the polity*".[21] The legitimate *insiders* (read citizens) being rewarded with the exclusive entitlements of citizenship.

Though state is only one among the many social organisations of human existence, fetishism with the geometry of border has ascended the entity of state into the Hegelian "master of space". This preoccupation with state and its sovereign limits has invariably resulted in what Buzan and Little identify as the "*Westphalian straitjacketing*"[22] of the discipline. When border is reduced to its material space, the disciplinary domain of international relations falls prey to what John Agnew calls as the "*territorial trap*".[23]

Till the onset of globalisation, this framework dominated the discipline. Globalisation and its advocates not only challenged the existential anchoring of the border but shook the very foundation of its master-concept, the state. Setting up of supranational organisations, international civil society organisations, birth of internet, global risks of climate change, epidemics, etc. underscored the increasing porosity of borders.

Contrary to the statist narrative of border, the spatial constitution of border demands a deeper and context-sensitive engagement. In the domain of international relations, the spectrum of engagement with the idea of space and therefore border largely oscillates between the polarities of "spatial fetishism" and "spatial exorcism".[24]

In other disciplinary domains, particularly like human geography and critical border studies, border discourse has effectively circumvented the "territorial trap". Pushing beyond the parochial understanding of border as geography of strategic significance, the space of border is understood more nuanced as "*site at and through which socio-spatial differences are communicated*".[25] Evolving from its preoccupation with border as a fixating territorial line, the post-modern turn in the discipline meant that border was to be approached as a human construct. This in turn marked a cardinal shift in approaching borders from "*political limits of states to socio-territorial constructs*".[26]

This paradigmatic shift from the bounds of "*methodological territorialism*"[27] meant a deeper engagement with the processes of border making aka *bordering,* thus widening the canvas of the border discourse.

The tapestry of border discourse is as diverse from the rigid understanding of border as the strategic power geometry to the acknowledgement of it as an ever-evolving "spatial practice" (Kolossov 2005). In a nutshell, there exists no singular metanarrative that explains, or say, defines the essence of border.

It is against this larger discursive backdrop that the Indo-China border of Tawang needs to be located.

## Locating Tawang in the India–China Border

The land boundary between India and China spans across 3488 km[28] (MEA) and it constitutes one of the enduring and intractable territorial disputes in the world today. Indo-China boundary contention can be located along three sectors: western, middle/central and eastern. In the west, Aksai Chin, a largely barren plateaus region with endorheic soda basins and a quintessential part of India's Union Territory of Ladakh, is claimed by China as part of its Xinjiang Uyghur Autonomous Region. Dispute over Aksai Chin has its origin in the British colonial cartography. Post-independence, India adopted the 1865 boundary line devised by British surveyor W. H. Johnson and that was later endorsed by the British General John Ardagh in 1897 in anticipation of the Russian foe. The legitimacy and historical accuracy of Johnson–Ardagh line is disputed by China. China's construction of Xinjiang-Tibet Highway in the 1950s and the encroachment of what India claimed to be its territory was one of the events that drew curtain to the Sino-Indian camaraderie, finally culminating in the 1962 Indo-China war.

Of the three sectors, middle or central sector extending from Demchok to border of Nepal, passing through the states of Himachal Pradesh and Uttar Pradesh is relatively tranquil. In 2002, India and China had exchanged their respective maps of the central sector.

With Sikkim incorporated into the Indian Union in 1975, boundary dispute in the eastern sector largely revolves around the state of Arunachal Pradesh, particularly the border district of Tawang. As is the case with the western sector, germplasm of the dispute happens to be the legitimacy of the boundary line drawn by the British diplomat Henry McMahon in 1914. Drawn along the Himalayan crest, the validity of McMahon line and the 1914 Simla Agreement is intensely disputed by China. Along with the British Indian and Tibetan counterparts, Ivan Chen, the Chinese plenipotentiary participated in the 1914 Simla Accord. However, China refrained from signing the Simla Agreement ratifying the McMahon line. The immediate reason being, China never recognised Tibet to be an independent country, therefore an unequal, with whom an international agreement could not be entered into. The twin disagreements over the colonial McMahon line and 1914 Simla Accord marks the genesis of the caustic eastern sector dispute.

Prior to boundary dispute taking an ugly turn in 1962, in 1960, Chinese leader Zhou Enlai had offered the one "package deal",[29] which China then believed to be a pragmatic resolution to a complex problem. Highlight of the deal was acceptance of the status quo, that is, India acknowledging Chinese claim over Aksai Chin and China accepting incorporation of Arunachal Pradesh/eastern sector into the Indian Territory. The deal was rejected by Nehru, as ceding to it would mean inefficacy of India in safeguarding its sovereignty. Interestingly, the 2017 visit of Dalai Lama to Tawang seems to have visibly reversed China's stand, reflected in the statement made by Chinese Special Representative Dai Binguuo, "*if the Indian side takes care of China's concerns in the eastern sector of their border, the Chinese side will respond accordingly and address India's concerns elsewhere*".[30]

At face value, it seems that the boundary deadlock between India and China pertains to the ambiguities and disagreement over colonial cartography and treaties. However, beneath the conflicting cartographic positions, post-independence approaches and imaginations have played a significant role in defining the trajectory of the territorial dispute. The destiny of independent India and China unfolded in similar settings, in 1947 and 1949, respectively. Both the countries inherited memories of a traumatic past. Centuries of British colonialism drained India of its economic fortunes, coloniser's divide and rule policy robbed India of its communal harmony. On the other hand, Opium Wars instigated by British imperial interests and a litany of Sino-Japanese battles severely punctured China's territorial and economic seam.

While post-colonial India decided to adopt the international boundaries set by the colonial administration, China embarked upon a different orbit, departing from the colonial cartography. The emotive motor force of Chinese nationalism proved to be the undoing of the miseries inflicted upon by the "century of humiliation".[31] To China, the boundaries demarcated by the imperial power were least acceptable, as adhering to it would mean psychological submission to the same. This was a significant decision as China shares its boundaries with no less than fourteen countries. This rejection of colonial boundaries was complemented by the Chinese commitment to bring all the former Chinese territories into the liberating fold of motherland. In his speech from the ramparts of Tiananmen Square on 1 October 1949, Mao Zedong announced the the ambitious political project of aggressive reunification. It is against this ideological backdrop, permeated with the emotion of national honour that Indo-China boundary dispute needs to be located.

Nitty-gritty of the colonial boundaries aside, as C Raja Mohan opines, "Tibet is the iron law that governs India–China relations".[32] Precisely for this reason, Tawang in Arunachal Pradesh stands at the heart of Sino-Indian territorial dispute. Perched in the western most part of the Indian state of Arunachal Pradesh, geographically, Tawang is bordered by Tibet Autonomous Region of the People's Republic of China in its north and the *druk* kingdom of Bhutan in its south-west.

Prior to the visit of the great Indian tantric master Padmasambhava (Guru/Lopon Rinpoche) in the eighth century, Tibet and its adjacent areas including Tawang largely remained under the influence of the animistic Bon religion. With the advent of Buddhism in the region and particularly after the construction of the majestic Galden Namgyal Lhatse aka Tawang Monastery by the fourth Merag Lama Lodre Gyatso, Tawang became one of the most venerated seats of Tibetan Buddhism. Ogyenling (Urgyelling) in Tawang happens to be the birthplace of the sixth Dalai Lama, Tsangyang Gyatso. The migratory blacknecked cranes, locally known as *thrung thrung karma*, which visit Tawang during winter are apotheosised as the reincarnation of the sixth Dalai Lama.

Till its incorporation into the Indian Union in 1951 by Major Bob Ralengnao Khathing, a Manipuri Tangkhul Naga frontier officer, and his troops, Tawang remained under the suzerainty of Tibet. Tax collectors called *dzongpons* from Tsona were stationed in the region, collecting tax from the people. Even today the *dzongs* (administrative centres designed like a fort) at Dirang, Senge and Talung, stand as reminders of the Tibetan past. Administration in the region was carried out by a six-member council called *trukdri,* which comprised of two *dzongpons* from Tsona[33] as well. British records of colonial expedition to the north of Ze La such as that of Captain Bailey and Morshead (1913) and the administrative reports of the officers of the Balipara Frontier Tract, like Captain Neville (1914), Captain G.S Lightfoot (1938) and J.S Mill chronicled the presence of Tibetan administration in the region. Their reports highlighted local miseries under a less charitable regime.

As far as the British were concerned, till the creation of McMahon line in 1914, Tawang largely remained *terra incognita.* Threat of Russian advance and Chinese belligerence kindled British interest in the region. The political developments from 1935 onwards forced the Frontier administration to take up the case of safeguarding the validity of McMahon line and if demanded, to assert British authority in areas of violation. As evident in "a letter dated 17th September 1936, from the Assam Chief Secretary to the Political Officer, Balipara Frontier Tract",[34] there emerged an imminent danger from the Chinese as many of "the latest maps published by China show almost the whole of tribal area south of the McMahon line up to the administered border of Assam as included in China. It amounts to this, that while the Chinese already claim a large stretch of Indian Territory east of Tawang as part of the Sikang province of China, the Tibetan government over whom the Chinese claim suzerainty, are collecting revenue and exercising jurisdiction in the Tawang area many miles south of the international frontier".[35]

With a relatively improved surveillance and sensitively designed administrative measures, British tacitly invoked the McMahon Line, reminding Tibet of its trespassing. Perhaps as a fallout of these measures, the1945 Report of J.P. Mills records the dwindling Tibetan presence in the region. However, owing to the economic pressures and contingencies caused by World War II, colonial administration's interference in the region was cautious, reluctant to execute the recommendations made by its officers of the Balipara Frontier Tract. This is particularly evident in the response to Captain G.S Lightfoot's recommendations in 1938. In 1947, when British left India, future of the unsettled case of McMahon Line was shelved to destiny.

In 1951, when Major Bob Khathing and his party evicted the Tibetan administration from the region, thus incorporating Tawang into the Indian state, China adopted a strategy of informed silence. However, with the 1959 episode of fourteenth Dalai Lama's exile to India via Khinzemane in Tawang, the region became a flashpoint in the Indo-China boundary dispute. Tawang's subjecthood to Tibetan suzerainty in the past and its thriving Monpa population, who are devout practitioners of Tibetan Buddhism became the rallying points for Chinese claim of Tawang. Nehruvian Forward Policy and the ambiguities pertaining to the McMahon Line led to the skirmish over India's setting up of Dhola post in the region, which China claimed to be on its side of the McMahon Line, eventually flaring up to the 1962 border war. Attesting the argument that countries with territorial disputes have high propensity to war[36], 1962 marked the lowest point in Indo-China relations.

A theatre to the 1962 war, Tawang remained captive for a brief spell of time till China voluntarily retreated the territory in November 1962. The 1987 Sumdorong Chu incident near the Thagla ridge became another confrontation of heightened tension. Routinely, visit of any Indian or foreign dignitaries in Tawang is reparteed with swift condemnation by China. The incident of China issuing stapled visas for Indian citizens from Arunachal Pradesh proved to be a classic case of rhetorical Chinese manoeuverings.[37] With China hastily damming the water of Yarlung Tsangpo (Brahmaputra) in Tibet, conflict over riparian resources has initiated India to consider construction of dams in Tawang. Coupled with strategic concerns, harnessing the surplus hydropower potential of the region resulted in Eighth National Hydropower Plan proposing the construction of of thirteen dams in the ecologically sensitive border district of Tawang.

From Tibetan suzerainty to colonial administration to Indian citizenship, Tawang has traversed rugged terrain of multiple subjecthood and border imaginations from above. These diverse imaginations of Tawang through various prisms of power share a common thread of power fuelled interests be it the strategic, national interests of the state or expansive agenda of the colonial great game or the power calculus of a theocracy. However, beyond these shifting conceptions of border, Tawang also happens to be an intimate space of existence for the Monpas of Tawang. To them, rivers of the region are not merely fluvial resources that need to be tapped for its monetary promise nor statements of strategic rivalry. Worshipped as sacred rivers and soul lakes, flowing water occupies a significant place in their religious pantheon, be it the prayer wheels turned by running water called *dungyur mani* or some of the funeral rituals that involves immersion of the corpse cut into an auspicious numeric of 108 pieces.

Besides, in a land stressed area like Tawang, increasing land acquisition for defence purposes by the Indian state is a matter of local concern as many of the land acquisitions mean a compromise on the available grazing land. Beyond the anthropocentric understanding of the region, Tawang is one of the biological treasure troves in the Indo-Burma hotspot region, home to rich varieties of flora and fauna.

In the national circuit of development, Tawang qualifies to be a dual periphery, both as the limits of India's sovereign power and a "resource frontier" in North East India,[38] bearing the brunt of development deficit. Even after seventy years of independence, Tawang remains one of the inaccessible borderlands of India, with perennially patchy roads. Majority of its ordinary population with critical illness does not make it beyond the S (Se La). With rare combination of favourable weather and luck, it takes a full day in reaching Guwahati or Shillong, this frontier district's nearest, advanced medical facilities.

The rich universe of space and meaning-making in and of Tawang cannot be reduced to the geopolitical significance of its territoriality alone. Geodesy of border is only one among the many identities it bears. As critical border theorists such as Olivier Kramsch and Ansi Paasi suggest, beyond the static trappings of the border nomenclature, the space has to be engaged with the dynamism of the verb "*bordering*" than the stasis of the noun border. Exploring its multiple geographies-moral, social and cultural would be rewarding if not therapeutic of the Westphalian "territorial trap".

## Conclusion

Tawang sits at the juncture of two memoryscapes. Spiritually seeking refuge in the pantheon of Tibetan Buddhism and thereby the authority of Dalai Lama, Tawang never insulated itself from Tibet. In the material realm, with Tibetan *dzongpons* (tax collectors) stationed in Tawang till the Indian expedition led by Major Khathing in the 1950s, it was at the receiving end of the Tibetan administrative set up. The 1962 Indo-China war drew iron curtain between Tawang and Tibet (Tibet Autonomous Region), making it one of India's hard borders. However, beyond the strategic logic of territory, the cultural and ethnic strings of identity still runs deep.

Guided by Deng Xiaoping's foreign policy prudence,"*hide your strength and bide your time*", post-independence China managed to 'resolve 17 out of 23 of its boundary disputes, in few cases renouncing parts of its territory' (Tweed 2018) as well. Hence, it seems the emotion and logic of Sino-Indian boundary deadlock runs deeper than the conventional configuration of a classic boundary dispute. Apart from the convoluted terrain of historical and documentary evidence, two factors that limit the possibility of territorial dispute settlement between the two Asian neighbours are the absence of unambiguous geographical demarcation and intense disagreement on the legitimacy of treaties, particularly the 1914 Simla Agreement. Given the complex annals of boundary disputes across the globe, these are not hurdles insurmountable on the road to peace. Then, what could be the algorithm that sustains the intransigence of the Indo-China boundary dispute, particularly in the case of Tawang?

Firstly, there exists incongruence between the Indian and Chinese post- independence imaginations of the geopolitical contours of the state. For instance, when India decided to carry forward the colonial cartographic legacy, except in the case of Pakistan, China was determined to denounce the colonial vestige and to reinstate "the imperial limits of the Qing dynasty".[39] This disjuncture would mean that the conventional principle of "*uti possidetis ita possidetis*" (as you possess, thus you may possess) invoked by the International Court of Justice may not pave path for reconciliation.[40]

Secondly, the nature of China's boundary disputes is intimately associated with its national imagination rooted in undoing the humiliation inflicted by its unpleasant past. Even in the event of border skirmishes, Chinese intent seems to be less inspired by territorial expansion. For instance, in the 1962 Indo-China war, occupation of Tawang by the Chinese army and its unilateral ceasefire and retreat had less to do with broadening of its territorial horizon. China's 1950 occupation of Tibet, what it calls "liberation of Tibet", needs to be read against this backdrop, as an effort to expunge the British policy of bifurcating Tibet into Inner and Outer Tibet.

A quaint periphery largely inhabited by the Buddhist Monpas who were erstwhile subjects of Tibetan suzerainty, to China, Tawang is critical in entrenching its logical primacy over the now Tibet Autonomous Region. Interestingly, in 2017, China has opened a museum in Lhasa commemorating the life of Tsangyang Gyatso, the sixth Dalai Lama, particularly his acclaimed mystic literature. Given the alleged Chinese hand in the abduction of Tsangyang Gyatso and his eventful death, this curious invigoration of the legacy of the sixth Dalai Lama is indeed in conjunction with China's assertion of Tawang.

Unabated non-military provocation by China, including the March 2019 destruction of 30,000 maps for inaccurate portrayal of its boundaries, including Tawang hints at the probable future trajectory of the Indo-China boundary dispute, at least in the Eastern Sector. For China, the very "intractability" of the boundary dispute is a carefully orchestrated one, going beyond the emotive element of national honour. One proposition that could better explain the rationale of maintaining an irreconcilable rivalry could be that of Krista Wiegand's argument of territorial dispute as a strategy of bargaining, or say a tool of bargaining leverage. In "*Enduring Territorial Disputes: Strategies of Bargaining, Coercive Diplomacy and Settlement*", Wiegand employs four case studies including Russo-China boundary dispute to make sense of what she calls "enduring disputes".[41] According to Wiegand, the perpetual sustenance of boundary disputes is nothing but an expression of bargaining, improving one's prospect in the volatile power hierarchies. This could well define the nature of the never-ending rival claims of Indo-China boundary disputes as well.

Apart from the claim over the riparian resources, which China sought to redirect to its parched geography and India to its power stressed domestic circuits, the real burden of border construct falls upon the Monpas of Tawang. With India and China competing to play the Tibet card to further their strategic interests, what goes unnoticed is the local resource conflict escalating between Tawang Monpas and the state apparatus over land, which could challenge the Indian narrative of this border in future, if unaddressed.

Amidst the chaotic power conundrum, for Tawang Monpas, the border construct subsumes alternate identities and possibilities of meaningful existence. The dehumanisation of the border population as "strategic assets" by the statist lexicon is clearly reflected in the engineered accessibility of the region to resources and particularly infrastructure which are critical for a dignified existence. A denial of the same could become a breeding site of unrest, as was evident during the 2016 anti-dam protests. A normative road map would imply a substantive shift from a state-centric, security driven imagination of space to a more human-centric one where understanding of self and the world around are not circumscribed by strategic narratives alone.

**Endnotes**

1. Nail (2016).
2. Paasi (2011).
3. Ratzel (1903).
4. Laine (2015).
5. Ibid.
6. Ibid.
7. Ibid.
8. Ibid.
9. Ibid.
10. Ibid.
11. Ibid.
12. Ibid.
13. Kristof and Ladis (1959).
14. Laine (2015).
15. Kolossov (2015).
16. Ibid.
17. Ibid.
18. Laine (2015).
19. Ibid.
20. Kolossov (2015).
21. Vaughan-Williams (2009).
22. Buzan and Little (2001).
23. Agnew (1994).
24. Strandsbjerg and Kleinschmidt (2010).a
25. Houtum (2005).
26. Ibid.

27. Schendel (2004).
28. Management of Indo-China Border.
29. Katherine (2015).
30. Anuja (2017).
31. Goswami (2012).
32. Joshi (2018).
33. Arpi (2013).
34. Reid (1942).
35. Ibid.
36. Vasquez (1995).
37. Goswami (2012).
38. Tsing (2003).
39. Noorani (2008).
40. Ibid.
41. Wiegand (2011).

## References

Agnew, John. 1994. The Territorial Trap: The Geographical Assumptions of International Theory. *Review of International Political Economy*.

Anuja, Atul. 2017. China Shifts Focus on "Eastern sector" As the Core of Border Row with India. *The Hindu,* dated 3rd March, 2017.

Arpi, Claude. 2013. *1962 and the McMahon Line Saga*. USA: Lancer Publishers.

Buzan, Barry, Richard Little. 2001. Why International Relations has Failed as an Intellectual Project and What to Do about It? *Millenium: Journal of International Studies* 30 (1): 19–39.

China Destroys 30,000 'Incorrect' World Maps, *The Hindu* dated 26th March, 2019.

Goswami, Namrata. 2012. China's Territorial Claims on Arunachal Pradesh; Alternative Scenarios 2032, IDSA Occasional Paper No. 29, Lancer's Books, New Delhi.

Hartshorne, Richard. 1950. The Functional Approach in Political Geography. *Annals of the Association of American Geographers* 40.

Havlicek, T., Jerabek M.,and Dokoupil, J. (ed). *Borders in Central Europe after the Schengen Agreement*. Springer

Houtum, H. van. 2005. The Geopolitics of Borders and Boundaries. *Geopolitics* 10.

Joshi, Manoj. 2002. India China Agree to Exchange Border Maps. *Times of India*, 31st March, 2002.

Joshi, Manoj. 2018. India's So-Called New Policy on Tibet is Neither New Nor Effective. https://scroll.in/article/826269/indias-so-called-new-policy-on-tibet-is-neither-new-nor-effective.

Katherine, Richards. 2015. *China-India: An Analysis of the Himalayan Territorial Dispute*. Source: https://www.defence.gov.au/ADC/Publications/IndoPac/Richards%20final%20IPSD%20paper.pdf.

Kolossov, Vladimir. 2005. Border Studies: Changing Perspectives and Theoretical Approaches. *Geopolitics* 10: 606–632.

Kolossov, A. Vladimir. 2015. Theoretical Approaches in the Study of Borders (Chapter 1.2). In *Introduction to Border Studies,* ed. V. Laine Sevastianov, Laine Sergei. P., Jussi and Kireev A. Anton, 33–56. Vladivostok: Dalnauka.

Kramsch, Olivier. 2010. Regulating European Borders: A Cultural Perspective. In *Innovative Regulatory Approaches: Coping with Scandinavian and European Policies*, ed. N. Veggeland. New York: Nova Science Publishers.

Kristof. D.K. Ladis. 1959 September. The Nature of Frontiers and Boundaries. *Annals of the Association of American Geographers*, 49 (3) (Part I): 269–282.

Laine, P. Jussi. 2015. A Historical View on the Study of Borders (Chapter 1.1). In *Introduction to Border Studies*,. ed. V. Laine Sevastianov, Laine Sergei, P. Jussi and Kireev A. Anton, 14–32. Vladivostok: Dalnauka.

Lefebvre, Henri. 1974. *La production de l'espace*. Paris: Anthropos.

MacKinder, Halford. 1904. The Geographical Pivot of History. *Geographi-cal Journal* 23: 421–44.

Mackinder, Harold J. 1996. *Democratic Ideals and Reality: A Study in the Politics of Reconstruction.* Washington D.C.

Management of Indo-China Border. *Ministry of External Affairs* https://mha.gov.in/sites/default/files/INDO%20CHINA_05052017.pdf.

Nail, Thomas. 2016. *Theory of the Border*. USA: Oxford University Press.

Newman, David. 1999. Boundaries, *Territory and Postmodernity*. Abingdon: Frank Cass (Taylor & Francis).

Noorani, A.G. 2008, October 24. Maps and Borders, *Frontline*.

Paasi, Anssi. 2005. Generations and the Development of Border Studies. *Geopolitics* 10: 663–671.

Paasi, Anssi. 2011. A 'Border theory': An Unattainable Dream or a Realistic Aim for Border Scholars? In *A Research Companion to Border Studies*, ed. Doris Wastl-Walter. Doris: Aldershot, Ashgate.

Ratzel, Friedrich. 1903. *Politische geographie: Oder die geographie der staaten, des verkehres und des krieges*. Oldenbourg: Munich.

Reid, Robert. 1942. *History of the Frontier Areas Bordering on Assam from 1883–1941*. Guwahati: Assam Government Press.

Schendel, Willems van. 2004. *The Bengal Borderland: Beyond State and Nation in South Asia.* Anthem Press.

Scott, James and Kolossov Vladimir. 2013. *Selected Conceptual Issues in Border Studies*. https://belgeo.revues.org/10532.

Strandsbjerg, Jeppe and Jochen Kleinschmidt. 2010. *After Critical Geopolitics: Why Spatial IR Theorizing Needs More Social Theory* at https://www.researchgate.net/publication/268802101_After_Critical_Geopolitics_Why_Spatial_IR_Theorizing_Needs_More_Social_Theory.

Tsing, Anna. 2003 Nov. 29–Dec. 5. Natural Resources and Capitalist Frontiers. *Economic and Political Weekly* 38 (48): 5100–5106.

Tweed, D. 2018. *China's Territorial Disputes*. Bloomberg. As updated on October 4, 2018. https://www.bloomberg.com/quicktake/territorialdisputes

Vasquez, John A. 1995. Why Do Neighbors Fight? Proximity, Interaction, or Territoriality. *Journal of Peace Research* 32 (3): 277–93.

Vaughan-Williams, Nick, and Border Politics. 2009. *The Limits of Sovereign Power*. Edinburgh: Edinburgh University Press.

Whittermore, S. 1940. *Boggs, International Boundaries: A Study of Boundary Functions and Problems*. New York: Columbia University Press.

Wiegand, E. 2011. *Krista, Enduring Territorial Disputes: Strategies of Bargaining*. Coercive Diplomacy and Settlement: University of Georgia Press.

**Sherin Ajin** is a Doctoral scholar at the School of Conflict and Security Studies, National Institute of Advanced Studies (NIAS), Bengaluru and has registered with the Manipal Academy of Higher Education (MAHE), Manipal, India in their PhD programme. Belonging to the disciplinary domain of Political Science, her areas of interest include Critical Border Studies, identity politics and discourses of state making and development.

# Chapter 3
# Tawang and the Mon in their Borderlands: A Historical Overview

**Neeru Nanda**

## Introduction

Tibetan Buddhist texts present Monyul as a territory lying just south of Tibet, bordered by the Himalayas on the north and the Assam foothills on the south. This territory was loosely referred as the Land of Mon or the lower land. On closer examination, however, definitions of Mon as a geographical locale turn out to be imprecise as several other tribes also inhabited this Himalayan belt and were termed as the Lhopas by the Tibetans texts, wherein they are described as the wild and intractable people. This appellation of Lhopa may be taken to cover all non-Buddhist tribes of Arunachal Pradesh, such as the Aka, Miji of Kameng district, Nishi, Tagin and Apatani of Subansiri, the Adi of Siang, Mishmis of Lohit.[1]

Lhopa territory generally meant trouble for the lamas, traders and pilgrims from Tibet. The Akas of West Kameng have had a long history of clashing with Tibetan traders and encroaching upon their traditional trade routes which extended from Poshingla in Tawang, southwards through Dirang, onto the foothills of Assam. On occasions, the Tibetans were forced to pay taxes and exactions from the local tribals![2] The proselytising efforts by the Buddhist monks could not penetrate into Lhopa land and the jurisdiction extended only over the Tawang Gompa. The lands of the Mon were reported to be always pleasant and welcoming to exiles fleeing Tibet due to persecution, pestilence and harsh vagaries of climate. Padma Lingpa, the Monpa patron saint of Bhutan describes it thus: "A time when earth and hail pound the crops, A time when the people of Tibet come forth to Mon".[3]

However, it should not be forgotten that Tibetan religious texts also regarded the Mon as inferior or even barbarians to whom the civilising influence of Buddhism was brought from Tibet.

---

N. Nanda (✉)
Tawang, India
e-mail: neeru.nanda2333@gmail.com



M. Mayilvaganan et al. (eds.), *Tawang, Monpas and Tibetan Buddhism in Transition*,
https://doi.org/10.1007/978-981-15-4346-3_3

## In Search of the Mon

The search for Mon begins with a focus on communities which figured as Mon, either in their self-view or in the Tibetan Buddhist texts. Tawang, Dirang and parts of Western Kameng of Arunachal Pradesh, were historically and territorially (traditionally) part of the "Monyul". The inhabitants are now classified as the Northern Monpas of Tawang, the Central Monpas of Dirang and the Southern Monpas of Western Kameng. A popular saying in Tawang describes themselves in simple, direct and proudly declaring that they are Mon: "We are neither of Gyaser nor of Gyakhar. We are the Mon".[4]

Before its unification in the seventeenth century, Bhutan was called Lho Mon Kha Bli and mentioned thus in the Tibetan religious texts, which may be traced to as early as in 1431.[5] It was viewed as a territory demarcated by four prominent boundary landmarks (kha-mouth, bli-four).

Today, the people of Bhutan call themselves Drukpa and their land as Druk Yul. In his excellent and erudite History of Bhutan, Karma Phuntsho points out why "Mon" is no longer acceptable to the nation-state of Bhutan as they have come to view the terminology as somewhat a derogative term coined for its inhabitants by the Tibetans.[6] Karma Phuntsho quotes Buddhist texts of Bhutan which present *Bo* (Bhot or Tibetan) and *Mon* as belonging to two separate geographical and cultural realms. So the people of Bhutan, (the Mon referred in the quoted religious texts), also affirmed a separate identity which they cherished and sought to protect, particularly vis-a-vis the identity of the Tibetans.[7]

The Monpas of West Kameng had been a part of Eastern Bhutan (SharMon) prior to the consolidation and unification of Bhutan in the seventeenth century. Early kings of Mon had ruled East Bhutan and West Kameng, as recorded by historian Micheal Aris.[8] The ruling clans, kings and chieftains had been based out of Domkho and Morshing which are now located in the West Kameng. These Southern Monpas retained their identity as the Mon right through British rule when they were recognised as one of the Sat Rajas of Aitchinson's Treaties.[9]

Captain Bailey's report of British forays into Tibet in 1914 details the first British records of a visit to Tawang in 1914.[10] Captain Bailey had independently grasped the truth of the self-image of Tawang Monpas. He astutely made the connection between Tawang and Bhutan. Thus he reports, "in the nature of the country they inhabit, their customs, language, dress and methods of building, the Monpas are very distinct from the Tibetans and resemble more the inhabitants of Bhutan and Sikkim". Captain Bailey was not only correct in guessing the affinity between Bhutan and Tawang (which has been discussed in the previous paras) but also right about the affinity of both to the Lepchas of Sikkim.

The Lepchas of present-day Sikkim, Darjeeling and Kalimpong, figure as Mon in the Tibetan Buddhist texts. The Tibetan exiles and Lamas colonised the lands of this indigenous tribe, sometime in the seventeenth century through the proselytising activity that succeeded but failed to impose their name on the Lepchas who continued to identify themselves as Rong, and not Mon. It was only under the British administration the Rong identified themselves as the Lepcha that continues until date.

It is interesting to note here Karma Phuntsho's mention of one Monpa by name Achug, from Sikkim, at whose instance the Bhutanese had undertaken forays into the Kingdom of Cooch Behar.[11] This Sikkim Monpa of Bhutanese text could be none other than a Lepcha of the British records!

## Mon Studies and Linguistic Affinities

Was there indeed at one time a race or communities which were indigenous to the Himalayas? It is a daunting question and begs concentrated research and scholastic endeavour. In travel across the Himalaya, from the Kali valley in Kumaon (Byans, Chaudas and Darma) of Uttarakhand in the west, up to the Lepchas of Sikkim, in the east one would come across intriguing common threads of language and culture which seem to be floating as dwindling straws in the wind, sending intimations (romantic and imagined, perhaps and, certainly, soon to vanish) of a gentle and civilised race that may have at some distant point in time, inhabited these tracts of the Eastern Himalayas.[12]

Some of the literatures dealt with this very interesting aspect. For instance, Murty has devoted a comprehensive chapter to the Mon in his book *Paths of Peace*.[13] This book is by far the best guide to Mon communities of the Himalayas where the records of their presence in the Trans-Himalayan and Central Asian region, extending up to the Mon of Burma, have been discussed.

The *Grammar of Lepchas* by Colonel Mainwaring in 1876[14] contains a perceptive and sympathetic study of the Lepchas and reveals how their language is one of the earliest Himalayan languages to have its own distinctive script, a sophisticated grammar and an extremely rich vocabulary. It was the state language in Sikkim up to the early days of British rule.

Apparently, Mainwaring was aware of the Mon of Burma and that Mon people are also found in Ladakh. These "Lepchas" of Ladakh would be the same minority communities of Ladakh identified as Mon by Franke in his *History of Western Tibet*.[15] Murty mentions interesting comparisons made by Franke between Mundari and the dialect of Zanskar (Ladakh). He extends this comparison to the dialects of Monpas of Kameng. For instance, both Mundari and Mon of Zanskar have a common numerical system based on counting in the twenties which is also found in Tawang.[16] Yet another striking feature of Lepcha language is found only in Sanskrit and Mundari viz the dual count, in addition to singular and plural. This is a mark of very ancient languages like Sanskrit and Ancient Greek. However, Mainwaring states that Lepcha language yields evidence of being unquestionably far anterior to the Hebrew or Sanskrit.[17] Since a modern grammar of the Lepcha language has recently been drawn up in the Brill's Series of Languages of the Great Himalayas it would be possible and interesting if this claim, made in all seriousness by Mainwaring, could be examined in the realm of comparative linguistics, (if only to be refuted!).

Though the Mon of Burma have been absorbed into the Northern Burmese linguistic and cultural matrix, the London School of African and Oriental Studies had at one time a department of Mon Studies devoted to the language and culture of the Mon of Burma. One of the signal achievements of that department was a compilation of *The Dictionary of Modern Spoken Mon* by Shorto.[18] This valuable work adds usefulness for subsequent studies of linguistic affinities of Himalayan communities calling themselves, either Mon or Rong.

Monpa and Tshangla of Bhutan are reported to be similar in preliminary studies based (once again!) on numerals.[19] So far there has been no in-depth linguistic comparison of dialects of the Monpas of Kameng with the rich and diverse dialects of Eastern Bhutan, let alone the distant dialects like Lepchas, and Zanskaris of Ladakh. However, significant documentation of individual languages of remote and endangered tribes and communities of the Great Himalaya is under the leadership of George van Driem whose research centre has published the *Grammar of Bumthang*, the heartland of Mon culture before unification of Bhutan and its sectarian separation from Tawang.[20]

## Tawang and Bhutan Before the Tawang Gompa

The history of Tawang does not begin with the Tawang Gompa which was constructed around 1680. The place name Tawang has appeared only after 1680.[21] It is now almost forgotten that just over three hundred years ago, Tawang, West Kameng and Bhutan were one socio-economic and cultural realm, existed without borders.

Prior to 1640, multifarious Buddhist sects peacefully co-existed. The Karmapas, Kargyupas and Nyngmapas predominated in Tawang and East Bhutan (Shar-Mon), while the Drukpa monasteries were well established in Western Bhutan. The monasteries following the Yellow Sect was yet to be established in Tawang.

In the Dakpanang area of Tawang, (lying just north of Tashigong in Bhutan), the ancient Bon religion held sway. It is not to be confused with the Bon of Tibet which is also a Lamaist religion like the Tibetan Buddhism. The religion of Bon is a living faith in this region, based on the first-hand accounts and direct interaction with villagers and Bon priests.[22] The worship of sacred peaks is central to this indigenous animistic faith and the tales of the Mountain Gods of Dakpanang and Khet Gyamdong make fascinating reading that shows how deeply the people are rooted in the sacred landscape which they have woven around themselves through magical belief that hardly depends on grand institutions of any particular religion or sect. It is for the scholars to see how far this common thread of an indigenous faith rooted in sacred geography unites many of the oldest communities in India, continuing in our midst as an invaluable living heritage.

The entire LhoMon area owed a great deal to the teachings of Tibetan Buddhism, but the Buddhist religion in Monyul followed its own path, unrelated to the ebb and flow of Buddhism in Tibet. One of the reasons for the difference could be that the Tawang Monpas were essentially organised as democratic, self-governing communities, called Mangma and were headed by Tsorgens who varied from being elected or semi-hereditary chieftains.[23]

Buddhism is generally believed to have spread in Tawang and Bhutan from the time of the advent of Guru Padmasambhava in the eighth century. Yet for eight hundred years after there were still no large monasteries in Tawang. The free and democratic structure of the society allowed all sects to flourish and was welcoming to all Gurus of all sects. This tolerance and multiculturalism extended through Bhutan and West Kameng up to Assam. The Bhutan Chronicles records visits of the kings of Kamata (in lower Assam) to Domkho Morshing (in West Kameng) for receiving blessings of Kargyupa Lamas who were the traditional Gurus honoured by the Kings/chieftains of SharMon who ruled from Domkho Morshing.[24]

Thangton Gyaltso, the great bridge-building Lama of the thirteenth century paid extended visits to Bhutan and Tawang where he is held in great veneration. Bhutanese texts mention his visits to Assam where he seems to have had devotees amongst the kings. Tibet on the other hand has tended to present a history of internal strife, largely based on religious and sectarian rivalry which at its most sordid, seems to have been in competition for the land, goods and serfs.

LhoMon thus became a safe haven for Tibetan Buddhists fleeing persecution of the Bon kings and Bon lamas of Tibet. The proof of continuity of Buddhism in Monyul lies in the large number of holy texts which were hidden in this area by the Tibetan exiles fleeing, not only to save their lives from Bon persecution but also to save the Buddhist religion. Pemalingpa, the great Monpa Lama and patron saint of Bhutan, discovered many such texts and also acquired a large number of texts in his travel across Bhutan, Tibet, Tawang, Kameng and Assam. He is the patron saint of Bhutan and was a Monpa with close family relations in Shyarcho in Tawang.[25]

Equally, Guru Padmasambhava's travels and teaching dot the countryside of Tawang and Bhutan. While he converted the land into a sacred geography for Buddhists, extending from Tawang to Bhutan, there is some textual evidence to indicate that he also visualised Tawang as a sacred landscape existing prior to his visits in the eighth century. Murty quotes from Du Kala Namdar, a Buddhist text of the Tawang Gompa, which describes Tawang, seen by Guru PadmaSambhava as he enters from the north. It is a riveting description of a sacred land, sanctified by saints who have meditated in various holy spots. The veracity of the description is borne out by the mention of place names and also by a description of the people and their clothes "the colour of flames stoked in the wind"[26]—a poetic rendition of the maroon colour of the distinctive Monpa dress obtained from the madder of Adina Cordifolia.

Pilgrimages followed ancient routes, to and fro, the sacred sites extending from Lhagaleng Gompa (now in Kameng) up to Domzang Gompa, just north of Tawang, (now in Tibet). Thangton Gyaltso, the great Chaksam (bridge-building) Lama of the thirteenth century further reinforced pilgrimage and trade routes connecting Assam, Bhutan, Tawang and Tibet. Huge amounts of iron ore and labour force were donated for these bridge-building projects by the villagers of Monyul (Bhutan).[27] In Tawang, Chaksam Lama built two crucial connecting bridges over the Nyamjang Chu and the Tawang Chu and many a Kakalinga is dedicated to him, with murals which incidentally depict him as an Indian Yogi.

Also, trade was a major source of well-being and wealth and the Mon people were not only just go-betweens or middlemen in this trade between Tibet and Assam, but they controlled trade in this region. Karma Phuntso quotes texts which describe the bounties of nature and the contribution of this region to the resultant flourishing trade. Food grains, medicinal herbs, sugu (daphne botanica) for the vast array of religious texts, even iron ore for bridge building in Tibet, were just some of the items (crucial to the Tibetan economy) which flowed northwards from this land of Mon. Tibetan medicine depended heavily upon the variety and profusion of medicinal plants available. The picture that emerges is that of a tolerant, pluralistic, rich and prosperous, dharmic, peace-loving and industrious society. The whole region was a vital source of material and human resource for Tibet, besides being the gateway to Assam.

## Tibetan Invasions of Bhutan

Closely related Mon people were partitioned purely as a result of sectarian strife within the realm of Gelukpa and Drukpa monastic orders of Tibet and Bhutan and Tawang in the seventeenth century. It was a partition that changed the entire geopolitical structure of the region with far-reaching implications for Tawang.

There used to be a vital trade link that had extended through Tawang and Bhutan (as they were then), being the shortest and easiest route available from Tibet to Assam. The route from Tsona dzong (Tibet) followed the Namkachu Valley (Tawang) over the bridge built by Chaksam lama, down to Tashigong (Bhutan) and Dewangiri (Assam). It is one of the few correct features shown on a topographical map of Bhootea Duars placed before the House of Commons in England in 1865. This time-honoured trade route was closed when the sectarian wars erupted in the late seventeenth century between Tibet and erstwhile Bhutan that remains closed since.

It has been mentioned earlier that Mon was the land to which exiles from Tibet fled as a result of persecution—whether of Bon kings or due to monastic rivalries amongst the Buddhists themselves. Ngawang Namgyal, the young Rimpoche of Ralung Gompa of the Drukpa sect in Tibet, was one such exile who had to flee Tibet around 1616. Since western Bhutan already had a number of Drukpa monasteries owing allegiance to Ralung Rimpoche, the latter could easily rally a strong force to face attacks from his rivals in Tibet. Roughly between 1636 and 1650, the fledgling state of Bhutan was attacked by the Tibetan rulers and later by the forces of the Fifth Dalai Lama. The Tibetan armies attacked Bhutan five times, between 1640 and 1686, only to suffer ignominious defeats. The Ralung Rimpoche (who came to be known as the Shabdrung Rimpoche in Bhutan) not only managed to emerge victorious but also consolidated Bhutan into a strong, unified nation-state and thus it has remained till date.[28]

Tibet would have been completely cut off from the plains of Assam when unified Bhutan closed its doors. Tawang was the chicken—neck, which could have completely blocked Tibet out of its resource-bank and vital trade routes. However, Tibetan forces were saved by the loyalty of the Tawang Monpas who chose to side with the Dalai Lama. Thus, Tawang Monpas forestalled Shabdrung Rimpoche's onward March north. The Drukpa forces had to call a halt at Tashigong where they built a fort (dzong) around the same time as the Tawang Gompa was constructed in Tawang.

Providential advance planning had been done for the Fifth Dalai Lama through a Gelukpa monk called Mera Lama. Mera Lama had built a large monastery in his native Mera Sakden area of present-day Bhutan, but he was chased out of his village by the advancing Drukpa forces. The escape route led to Tawang where he was welcomed and given refuge. He rallied the Tawang Monpas, and soon converted them into followers of the Dalai Lama and the Yellow sect. The Tawang Gompa was finally built as a bastion of the Dalai Lama's Yellow Sect jurisdiction in the region. The date of construction of Tawang Gompa is generally placed between 1680 and 1686. Swords and shields which adorn the walls of the anterooms in Tawang Gompa bear mute testimony to the bloody battles between the Drukpas of Bhutan and the Gelugpa Monpa lamas who led the offensive-defence, being in the forefront of Tibetan forces in Bhutan and Tawang.[29]

Why did the Tawang Monpas side with the Fifth Dalai Lama of Tibet and not with the Shabdrung Rimpoche of Bhutan? After all, both the High Lamas were Tibetan, both were martial and intolerant of political and religious rivals. Further, the cultural and linguistic affiliations of Tawang Monpas lay with Bhutan?

The reason lies in the fact that the eastward March of the Shabdrung Rimpoche threatened to overwhelm political, religious, cultural and linguistic hegemony of the Sharchokpa Monpas of Eastern Bhutan. In fact the Drukpa hegemony of Western Bhutan was apparently too close for comfort. All the clans and chiefs of SharMon had to swear allegiance to the Drukpa sect and submit to the control of a unified administration.

The apprehensions of Tawang Monpas have been proved correct is borne out by present-day historical accounts of Bhutan which do not deny that LhoMon is the ancient historical name of their country, but still they effectively ignore/deny or "demonise" the existence of any kind of original Mon race. In fact, the Monpas of the remote pocket of Mera Sakden in Bhutan are officially categorised as Brokpas (graziers) and not as Monpa. Not one of the many dialects and languages of Bhutan identified by Gregory Van Driem are termed "Monpa" even though some of the major dialects (ex. Tshangla) are directly akin to Monpa language.

In hindsight, therefore, it would seem that the Tawang Monpas possibly made the best choice available to them for preserving their village institutions and their cultural identity as Monpas. Affiliation with the Lhasa regime at that time would have seemed remote and less formidable than the adjacent advancing forces of the

Drukpas. Thus, the Monpa language, ethos and identity were preserved in exchange for submission to Tibetan monastic jurisdiction over the Tawang Gompa which was made an affiliate of the Gompa just across the border in Tsona Dzong.

Having lost direct access to the Assam plains, but having managed to retain their hold over the chicken neck of Tawang, the first task of the Lamas heading the newly established Tawang Gompa became that of opening up an alternative trade route to Assam via the difficult terrain of Se La Pass. This route continues as the lifeline till today. An outpost at Dirang was later set up as a bulwark before the final journey on to trading outposts of Assam foothills. Thus were trade links of Tibet with Assam preserved. The cost was paid by the people on both sides of the borders of Bhutan and Tawang. Tawang was a major loser since the easy access to Assam plains via Tashigong was completely closed.

So a unified political entity, later called Bhutan or Druk Yul, emerged, hostile to its past connections with Tibet, and cut off from adjoining Tawang and the Gelukpa regime of Tibet. Tawang also slowly lost its original name of Tana Mandrekhang and came to be known as Tawang, even though the earlier surveys conducted in 1873 and 1876, mentioned the name of Tana or Mantangong.[30] Monyul was now effectively split. New religious and political identities emerged. But Tawang had yet to charter its own course.

## Monpas and the British Government

While the Monpas remained as self-governing communities and Tawang Gompa was very powerful, being overwhelmingly administered by Monpa Lamas, they seem to have realised, over time that they had been short-changed in the bargain with the Dalai Lamas regime.

The discovery of the sixth Dalai Lama amongst the Tawang Monpas could have secured the Tibetan-Monpa ties to a great extent, but the fact that he was a reluctant spiritual acolyte coupled with his untimely and mysterious death, forestalled any such long-term development. In the aftermath of rule by successive regents, the exploitation by Tibetan lamas and petty tax collectors only increased. They did nothing to endear the Tibetans to the Tawang Monpas. Though the latter were firm in their devotion to the institution of the Dalai Lama, (and remain so till date) the burden of demands for free porterage by Tibetan lamas and officials who travelled regularly collecting medicinal herbs, proceeding southwards for trade, etc. became increasingly difficult to bear. In addition, they had to face petty harassment by way of rivalries in trade and corruption of local officers.

It must be noted that the Khrai or grain tax paid collectively by each village to the Tawang Gompa, (albeit as per a "sanad" issued by the Fifth Dalai Lama) is actually a voluntary support extended by villages to lamas who were drawn from Tawang (the middle one of three sons always joined the monastery). It was never a bone of contention or grievance. It was a religious obligation even though it might lend itself to interpretation of being a tax imposed by Tibetan Secular rule on Monpa subjects. That this interpretation is far from the truth is borne out by the fact that Khrai has continued after 1951, when the Indian government took over administration of Tawang, and continues to be paid by the villages for upkeep of the Tawang Gompa.

Hence, the Monpas sought to free themselves as soon as an opportunity presented. This opportunity presented itself in the form of British forays and expeditions.

From 1913 onwards, the British records speak continuously of exploitation of Tawang Monpas and their discontent voiced before successive survey parties sent into the region by the British government.

In May 1914, a few months after Capt. Bailey's tour, narrated earlier, Tawang was visited by Capt G.A. Nevill who then went on to attend the Simla conference where he participated in drawing up the McMahon line. The World War meanwhile intervened and administrative expansion picked up thereafter only in 1936 and 1938 with the expeditions of Capt. Lightfoot who was deputed for the purpose of demarcation of the boundaries of Kameng and Tawang with Bhutan right up to the tri-junction with Tibet. He was also tasked with ascertaining the conditions under which the Monpas were living (in effect seeking to ascertain the revenue possibilities and other issues pertaining to gradual extension of active administrative jurisdiction to the Monpa areas). Capt. Lightfoot noted unequivocally that the Monpas of Tawang were a peaceful and friendly people—"the Tibetans they loathe and fear". The reasons cited were as noted earlier—forced labour and extortion of supplies, along with monopolist rates for salt (imported from Tibet) and rice (imported from Assam) occasioned by the fact that the entire trade route was in the grip of Tibetan traders and Tibetan tax collection agents at Dirang and Tawang.

The south of Se La, the Dirang Monpas, also had to face oppression and exactions not only from Tibetan traders but also from the Akas who had long controlled an ancient trade route to Tibet via Poshing La. It must be recalled that the Dirang Monpas had not actively sided with the Fifth Dalai Lama in the wars with Bhutan, nor participated in founding of the Tawang Gompa. They were, in fact, victims of the ambitious extension of Tibetan lama jurisdiction for purposes of trade, and therefore, it is but natural that Dirang would be the first to free itself from the yoke of trade control and taxes imposed by the Tibetan lamas of Tsonadzong.

By 1943–44, the British administration had started extending itself to the establishment of regular administrative outposts in the Himalayan belt south of the McMahon line. Hence began the second period of exploration and development in the Assam hills and its frontiers under the leadership of JP Mills, a senior Indian Civil Service officer with an unrivalled knowledge of the tribes of the area. As a follow-up of the various surveys and expeditions sent in from 1913 onwards, outposts were established in three prongs along the Lohit valley, the Siang valley and Dirang Dzong tract. Separately, Dr. Furer Haimendorf (accompanied by his wife) opened up the Apatani Valley.[31] Lt. Col. Betts, the political officer of the newly notified Balipara Frontier Tract, was assigned the task of opening up the Nishi and Miri belts of Subansiri district.

Subsequently, the Assam government had set up permanent outposts in Rupa in 1941 and in Dirang in early 1944. JP Mills, then Adviser to the Governor of Assam, himself visited Dirang in May 1945 and met with the Tsona Dzongpens to impress upon them, the legal validity and determination of the British government to put a stop to all exactions and extortions by Tibetan traders and lamas. Thus, ended the infamous Tibetan tax regime suffered by Dirang Monpas, south of Se La.

Almost a year later, Lt. Col. Betts toured the Se La Axis (as it was then called.) Their journeys have been described in vivid detail by his wife, Ursula Graham Bower, in her book *The Hidden land*.[32] The Agent Achung La, stationed at Dirang, was a charming and able Sikkimese to whom both political officer and his wife took a great liking! The Dirang Monpas, on their part, expressed profuse gratitude to the political officer for having restored their lives to a peaceful and settled condition with the setting up of outposts, freedom from forced porterage, the presence of medical outposts and security from exactions.[33]

## Tawang Monpas Greet Major Bob Khating

The Tawang Monpas must have been waiting in the wings, anxiously watching developments south of Se La. The complaints voiced by them before Capt. Lightfoot had echoed up to Shillong and had certainly made an impact on the British government officials in Assam. But they had to wait a little longer for permanent administration to move in. This took place soon after independence, when the leading Tawang elders assembled eagerly to greet Major Bob Khating as he marched into Tawang on 6 February 1951 and raised the Indian flag on 9 February 1951.[34]

The Indian authority, with its simple show of authority and determination, not least due to the forceful personality of the legendary Bob Khating, made it amply clear to the handful of representatives of Tsona Dzong that they would have to yield and comply with the Treaties of the Simla Convention 1914. The overwhelming majority of Tawang Gompa lamas were in any case local Monpas, totally supported by and dependent on the local villagers whose sympathies were clearly not with the Tibetans.

After observing Major Khating's activities for some time, the leaders approached him and requested for imposition of a nominal house tax. A government that did not impose tax would be viewed by them as an uncertain government, undecided whether to leave or stay. Major Khating, himself an intrepid Manipuri Naga officer, grasped the implications and agreed to impose the nominal annual house tax of Indian Rs. 5/- per household, which continued thereafter.[35]

The goodwill and trust of Monpas were earned very fast, by the wise and pragmatic frontier officers who served in the newly set up Northeast Frontier Agency. The loyalty of the Monpas and their willing aid to the beleaguered Indian army in 1962 has been certified. By all accounts, Monpas loyalties stayed with the Indian government despite the Indian army debacle and in spite of the deft Chinese propaganda unleashed during their two months of occupation. Though the Chinese were careful to avoid any kind of exactions or extortions, even their best behaviour could not make up for their attack on the Potala and subsequent flight of Dalai Lama to India through Tawang.

The Monpas have always been cautious and discriminating in making their choices. They were not going to cast their lot with the Chinese government which had destroyed the religious foundations of their society. On the other hand, the Tawang Monpas have also adroitly managed to discern between secular and sacred needs and compulsions. They would not translate their devotion to His Holiness into an overarching welcome to Tibetan refugees. Significantly, no Tibetan refugees (post 1962) were allowed to be settled in Kameng.

The greatest demonstration of spontaneous support of Tawang Monpas to the Indian frontier administration came with voluntary labour mobilised by the entire district for construction of 44 km of the Tawang–Lumla road in record time in 1977. An independent district for Tawang soon followed and Tawang leadership proudly contributed to representation in the council of ministers in Itanagar, crowned by the chief ministership of Arunachal Pradesh, which has been piloted in successive tenures by Dorjee Khandu of Tawang. The present chief minister of Arunachal, Pema Khandu is his son.

## As History Repeats Itself

Tawang has thus come on top priority for the Government of India and may be said to have attained a certain glory. But all that has come at a price. Now the local people are facing extortion of a very different kind—a disturbance which does not carry a face but threatens to uproot their eco-system and habitat in the form of construction of mega-dams across the Tawang Chu River. The local movement against these dams in Tawang was led by Monpa lamas of the Tawang Gompa, culminating in firing by the police in May 2016, which resulted in death of two lamas.

This utterly painful and astonishing incident and the movement behind it reveal some historical consistencies. The Tawang Gompa lamas have, at various points of time, played a decisive role in steering the course of history in the entire LhoMon region. Without the martial intervention and support of Tawang Monpa lamas, the Tibetan armies could never have defeated the Drukpa forces which were advancing on Tawang in the seventeenth century. The same Tawang Gompa lamas had finally settled matters in the wars between Southern Monpas and Sherdukpens in 1864–72. At that time, the legendary Monpa Lama Gelong Lobsang Tapgey had concluded a truce after the Tsona Dgongpens and High Lamas sent by Lhasa had failed to arbitrate between the warring factions.

Again, in 1908, the Tibetan authorities (no less than the Tsona Dzongpen himself) had to beat a hasty retreat from the precincts of the Tawang Gompa, even though accompanied by an armed force of Tibetans. It must also be remembered that when Major Bob Khating held his parleys with the Tibetan representatives of Tsona Dzong in February 1951, the silent overwhelming presence of the Monpa lamas was a force to be reckoned with—in the case of direct confrontation with the Indian government, their support (or lack of it) must have been seen by the Tibetan officers as a contending and crucial factor.

It is not surprising therefore, that the agitation against the hydro-projects and virtually against their own chief minister is being led by the Tawang Gompa Lamas. The Tawang Lamas do not view themselves as being solely inspired by the Dalai Lama's green narrative in their struggle to protect their fragile eco-system and dwindling population. The dam projects with their huge labour requirement and destruction of arable land and forests would be seen as leading to virtual extinction of the Tawang Monpas themselves, let alone their land and culture. However, it is useful to see how studying in the Sera Monastery in Karnataka has helped the young Lamas to hone their communication and networking skills and also to seamlessly integrate their struggle and project it onto a larger canvas. Hence, the analysis of linkages with the larger world external to Tawang, as made by Gohain, has significance and validity for all who seek to study recent socio-political developments in Tawang.[36]

Should the pro-dam establishment be viewed as impelled solely by utopian images of the developed world? Perhaps not entirely! No doubt the northeast is equally propelled by the consumerist vision of society which has emerged as a major discourse in post-liberalisation India, most evident from the change in themes and values projected by Bollywood and the whole gamut of electronic and other media. No doubt the leaders who profit from the democratic process and its attendant administrative structure are directly driven by images of this new society. But what holds the whole fabric together is the need to retain Power, for the leader, and the compulsions of closely knit societies to find new cohesive structures where the old no longer coincide with the new structures of power and pelf.

The result is a splintering process that has led to vertical splits within many of the traditional communities as they existed in the mid-seventies. Manipulations to hold on to a community of followers still continue to be articulated in terms of traditional family and village bonding centred on an established leader.

Interestingly, other attempts to oppose the chief ministers family by participating in elections have also been led by Lamas as in the case of the rather tragic figure of the Gompatse Rimpoche. He later committed suicide after having spearheaded a relatively unsuccessful movement to isolate Tawang, Central and Southern Monpa areas from exploitative and fast-paced development by bringing them under the Fifth Schedule. The full details and implications of this movement for an autonomous Mon region have yet to be documented.

An analysis of the inter-action between traditional hierarchies, social networks and the democratic processes and development strategies, imposed on remote self-governing communities, is perhaps the need of the hour.

Murty's contemporary, Rashid Yusuf Ali, also of the Indian Frontier Administrative Service, was another outstanding and insightful intellect, whose monograph *The Mandate of the People*[37] provides an excellent account of how traditional leaders of different tribes came to be included in the mainstream of political representation and administrative posts after independence. It documents a smooth transition into the new order of democratically elected government up to 1978 up to the tenure of Pema Khandu, who incidentally, belonged to West Kameng, and was the first chief minister of the Arunachal Pradesh.

Rashid Yusuf Ali has proved right in his assessment of the value of traditional leadership in governing the immediate course of future politics, but his idealistic vision of an emerging tribal democracy based on tribal traditions of consultation and consensus was nowhere near the hard realities of politics of the border state. The ground analysis of the course of politics in Tawang, seems to have been better summarised by Nanda:

"Tradition in the form of clan and family affiliations continued to play a strong determining role in the choice of candidates—in the political game, whether it is played in Tawang or Delhi, however, power, presents and patronage is the rule. Thus has Tawang been drawn into the mainstream of the nation?"[38]

## Conclusion

The Monpas being a part of the Indian polity safeguards this cultural identity along with their religious identity as Gelukpa Buddhists.

They are secure in their mother tongue and have notable linguistic skills leading to proficiency in Hindi and English. They have recently introduced study of Bodhi or Tibetan language because of their ardent Buddhist faith and the desire to be conversant with texts which shed light, not just on rituals and beliefs, but also on the history of the region.

There are straws in the wind which indicate a further interesting trend—a new vision of cultural revival. Local cultural organizations set up by educated Monpas have started documenting the Bon heritage of Tawang, along with the legends of Tana Mandre Khang and Gyebo Kala Wangphu which antedate the establishment of the Tawang Gompa. In an attempt to recreate the present through the past, workshops and seminars have been held to forge inter-linkages Himalayan communities, with affinity to the Mon, extending from Ladakh to Lepchas of Kalimpong, with a common interest in exploring and documenting their cultural heritage along with its pre-Buddhist elements.

There have been some important moves[39] towards bringing together the Pan-Himalayan Buddhist congregations of the Indian monastic orders—an endeavour which has been taken forward with the blessings of His Holiness. The Dalai Lama has reportedly christened it as "the Nalanda Tradition", in order to emphasise the uniqueness of the Indian Buddhist orders of the Himalayas.

Claude Arpi's interesting analysis of the Nehruvian approach that led to the unique philosophy of NEFA shows one of the consequences of isolating the people from the mainland has been this ignorance of the rest of India and a certain inability, born out of lack of contact, amongst Indians to view and accept the individuality of the people in the northeast. Arpi has perhaps put his finger on it by terming the Philosophy for NEFA as being driven by certain romanticism.[40] No doubt, it is the philosophy for NEFA which enables the young Monpa to confidently state—"We are neither the people of Gyaser (Tibet) nor the people of Gyakhar (India). We are the Mon". Being unique within the Indian polity helps them adjust comfortably within the many identities that India represents. The Monpas need no further integration. But the same philosophy also actively kept NEFA and the rest of India apart. The time for a holistic perspective of the northeast beyond the preaching for integration has come.

## Endnotes

1. Tibetan descriptions of the Lhopa of course, follow the Tibetan world view of superiority. Just one example-that of the superb and sophisticated cultivation of the Apatani valley-can bust this myth of wild tribes of Arunacahal. (Von Furer-Haimendorf 1962).
2. The Tagin Na and Mara tribals of Upper Subansiri used to exact taxes from Tibetan pilgrims undertaking the difficult pilgrimage of Tsari Che in the now disputed Longju sector. See Murty (1983), pp. 158–160. For more details, refer Arpi (2016a).
3. See Michael (1980).
4. The Gyaser, yellow or golden land, denotes Tibet which is the land of lamas and Gyakhar denotes the 'Shining' or 'White' land of Buddha and Divine Wisdom, India. A popular saying, recalled from the writers tenure in Tawang and recently reconfirmed from Maling Gombu, The Director, *Tawang Foundation*.
5. See Michael (1980).
6. See Phuntsho (2013).
7. See Ibid, p. 196.
8. For a dense but accurate account of the history of the Kings of Mon as related in Bhutanese texts and evidence of their continuation in Arunachal West Kameng. Refer Michael (1980).
9. See Michael (1980).
10. See Bailey (1957).
11. See Phuntsho (2013).
12. See Nanda (1992).
13. See Murthy (1983).
14. See Mainwaring (1971).
15. See Franke (1907).
16. See Murthy (1983).
17. See Mainwaring 1971.
18. See Shorto (1962).
19. See Michael (1980).
20. The dialect of Mon Bumthang would be very much akin to the Monpa dialects of Tawang and Kameng. The state government of Arunachal Pradesh needs to learn a lesson from pro-active Bhutan in this respect. (It is a pity that this particular publication seems to be available only in German). Refer Michael (1980).
21. See Michael (1983).
22. See Nanda (1980).

23. See Nanda (1980).
24. See Michael (1980).
25. See Michael (1988).
26. For the entire text of this fascinating description, see Murthy (1983).
27. See Phuntsho (2013).
28. See Phuntsho (2013). Also see Michael (1980), Part Three, Chap. 1.
29. Nanda, records a poignant story regarding the Sarong Rimpoche of Tawang Gompa as he returns with his bloodied sword from military frays in Bhutan. Refer, Nanda (1980), pp. 72.
30. For a detailed discussion on Tawang, Tana, Mandrekhang and Mantangong as place names found in earlier traveller accounts and religious documents, See Murty (1983).
31. See Von Furer-Haimendorf (1962).
32. See Bower (1978).
33. See Ibid, p. 145.
34. The feelings of leading Gaon Budhas (as at that time) have been expressed to me (author) as well as to other researchers—in short, they had felt let down by the earlier British expeditions because they had to suffer for the support extended by them to Capt. Lightfoot in the hope of being freed from Tibetan exactions. This time viz on 06th February 1951 they were even more anxious. Well, with Major Bob Khating and his posse of 200 Assam Rifles personnel, they did not have to wait.
35. It is a matter of interest that Ursula Graham Bowers, political officer wife, should also record how the villagers of Rupa and Dirang paid Indian Rs. 5/- in house tax annually, an amount they had chosen themselves (Bower 1978, pp. 145). Her account exactly corroborates the story related to me by Tawang elders about their dialogue with Major Bob Khating in February1951. Refer Nanda (1980).
36. See Gohain (2018).
37. See Ali (1978).
38. See Nanda (1980).
39. Two days Pan Himalayan Buddhist Conference that held in Gurgaon, 29 June 2018. See https://www.dailyexcelsior.com/2-day-national-conference-buddhist-culture-preservation-identity-begins/.
40. See Arpi (2016a).

## References

Ali, Yusuf. 1978. *Mandate of the People*. Itanagar, Arunachal Pradesh.

Arpi, Claude. 2015. *The Man Who Brought Tawang Under India*. The Pioneer, September 10, 2015.

Arpi, Claude. 2016. The Indian Frontier Administrative Service: Romanticism and Hostile Borders. In *Himalayan Bridge*, ed. Niraj Kumar et al., New Delhi: KW Publishers.

Arpi, Claude. 2016. *Getting a Grip on Issues Regarding Forward Areas*. The Pioneer, November 3, 2016.
Bailey, F.M. 1957. *No Passport to Tibet, Rupert Hart-Davis*.
Bower, Ursula Graham. 1978. New Delhi: Allied Publishers Pvt. Ltd. Reprint..
Franke, A.H. 1907. *A History of Western Tibet*. London.
Gohain, Swargajyoti. 2017. *Embattled Frontiers and Emerging Spaces: Transformatiion of the Tawang Border*, Economic and Political Weekly, April 15, 2017.
Gohain, Swargajyoti. 2018. Robes, Rivers and Ruptured Spaces. In: *A Place of Relations*, ed. Yasmin Saikia and Amit Baishya, Cambridge University Press.
Mainwaring, G.B. 1971. *A Grammar of the Rong (Lepcha) Language Bibliotheca Himalayica Series II*, vol. 5, Manjushri Publishing house.
Michael, Aris. 1980. *Bhutan, The Early History of a Himalayan Kingdom*. Delhi: Vikas.
Michael, Aris. 1988. *Secret Lives and Hidden Treasures*. Shimla: Institute of Advanced Studies.
Murthy, T.S. 1983. *Paths of Peace*. Delhi: ABC Publishing House.
Nanda, Neeru. 1980. *Tawang the Land of Mon*. Delhi: Vikas.
Nanda. 1992. In Search of the Rong. In: *Himalayan Heritage*, vol. I, Sundaram magazine, North Central Zone Cultural Centre Allahabad.
Phuntsho, Karma. 2013. *History of Bhutan, Random House*.
Sarkar, Niranjan. 1978. *A Historical Account of Tawang Monastery*. Shillong: Resarun.
Shorto, H.L. 1962. *A Dictionary of Spoken Mon*, Oxford University Press.
Sinha, A.C. 2001. *Himalayan Kingdom Bhutan, Indus Delhi*.
Van Driem, George. 2016. From Dhaulagiri to Lappland. In: *Himalayan Bridge*, eds. Niraj Kumar and George van Driem et al., KW Publishers Delhi.
Von Furer-Haimendorf, Christoph. 1962. *The Apatanis and Their Neighbours*, Routledge Kegan Paul.

**Neeru Nanda** an IAS (Retd.) and a former Additional Deputy Commissioner at Tawang. She has extensively written about Tawang, Monpas and Environment related issues. Her well know book is*Tawang, the Land of Mon*, published by Vikas Publishing House, 1982.

# Chapter 4
# Standing at the Himalayan Crossroads

## A Tale of the Liberation of Tawang

**Claude Arpi**

## Introduction

On 7 October 1950, the People's Liberation Army crossed the Upper Yangtze and invaded Tibet. Ten days later, Chamdo, the capital of Kham Province fell to the Communists. Eastern Tibet had been "liberated", proclaimed China. This happening would have grave consequences for the Himalayan border between India and Tibet. India had to wake up to the new realities. On November 7, in his letter to the Prime Minister, Sardar Patel listed some of the problems which, according to him, required "early solution and around which we have to build our administrative or military policies".[1] To cite a few, the Deputy Prime Minister mentioned:

- Military and Intelligence appreciation of the Chinese threat to India both on the frontier and internal security.
- An examination of military position and such redisposition of our forces as might be necessary, particularly with the idea of guarding important routes or areas which are likely to be the subject of dispute.
- An appraisement of strength of our forces and if necessary, reconsideration of our retrenchment plans for the Army in the light of the new threat. A long-term consideration of India's defence needs.
- The question of Chinese entry into UNO. In view of rebuff which China has given us and the method which it has followed in dealing with Tibet, I am doubtful whether we can advocate its claims any longer.
- The political and administrative steps which we should take to strengthen our Northern and North-Eastern frontier.
- The policies in regards to McMahon Line.

C. Arpi (✉)
United Service Institution of India, New Delhi, India
e-mail: claudearpi@gmail.com



M. Mayilvaganan et al. (eds.), *Tawang, Monpas and Tibetan Buddhism in Transition*,
https://doi.org/10.1007/978-981-15-4346-3_4

Nehru did not answer these questions for the simple reason that he already had been decided to safeguard "India's friendship" with China.

On November 20, the Prime Minister wrote to his ambassador in Beijing, "Our present policy is primarily based on avoidance of world war; and secondly on maintenance of honourable and peaceful relations with China. These relations inevitably will depend, to some extent, upon Chinese policy in Tibet. If peaceful settlement is arrived at there and Tibet's autonomy recognised, this should meet Chinese demands and satisfy, more or less, both Tibet and India".[2] Once the Chinese troops had stopped their advance in Chamdo for the winter, China was again considered as "peaceful" and Tibet was forgotten ...with it the vital consequences for India's borders.

## A Prelude to the Takeover of Tawang

### *The Bad News from Shillong*

A few days earlier, on November 12, Jairamdas Daulatram, the Governor of Assam, who was looking after the North-East Frontier Agency (NEFA) (the border with Tibet in the north-east) informed Indian Prime Minister:

> Have received wireless message from Assam Rifles Commandant at Sadiya[3] that according to message from Walong[4] issued 11th evening information has been received from Dak (postal) Runner from mules that arrived at Drowa Gompa[5] which is about 80 miles of North West of Rima,[6] which itself is 8 miles outside our own border. Two Tibetan Officers who had gone back from Rima were captured by these Communists. Assam Rifles patrol has been deputed to border for further enquires.

Less than a month after the fall of Chamdo, the Communist troops were fast moving towards the Indian border (i.e. the McMahon line). Daulatram told the Prime Minister that steps were being taken "to strengthen Walong outpost". He also told Delhi that Special Intelligence Staffs were leaving for Sadiya, "Also trying to ascertain from Gangtok (from Harishwar Dayal, the Political Officer) latest position as known to our Representative there".

Two days later, Sardar Patel sent a cable to the Governor bringing to his notice the dispatch (by air) of a company of regular Army troops to Agartala. Patel discussed the redisposition of forces, mentioned in his letter to Nehru on November 7, Daulatram was requested to send an adequate number of Assam Rifle platoons to the forward posts; they would be replaced by Tripura Rifles. The situation was considered extremely serious. At that time, the Government of India decided to form a Committee, known as the Himmatsinghji Committee. It was unfortunate that the findings of the Committee have now been "lost".[7] Besides Major General Himmatsinghji, Deputy Minister of Defence and Chairman, it also included Lt Gen Kulwant Singh, K Zakaria, Head of the Historical Division of the Ministry of External Affairs, SN Haksar, Joint Secretary, Ministry of External Affairs, Gp Capt MS Chaturvedi from the Indian Air Force and Waryam Singh, Deputy Director of the Intelligence Bureau.

In his *My Years with Nehru, China's Betrayal*, BN Mullik, the Director of the Intelligence Bureau (DIB) wrote that the decision to form a Committee followed a note entitled "New Problems of Internal Security" sent by him, which had been considered "by all the Ministries concerned within the next seven days".[8] Though Mullik liked to take credit for many things on behalf of the IB (except for the latter's failures), it was undoubtedly Sardar Patel who initiated the process and triggered the formation of the Committee. According to Mullik, two main decisions were taken, they are as follows:

- A small Committee of military experts with a representative of the IB based in Shillong would visit the NEFA agencies and propose the places near the frontier at which the Assam Rifles units should be posted.
- A high-powered Committee presided over by the Deputy Minister of Defence, Major General Himmatsinghji, with representatives of Defence, Communication, Home, External Affairs and the IB would be formed to study the problems created by the Chinese aggression in Tibet and to make recommendations about the measures that should be taken to improve administration, defence, communication, etc., of all the frontier areas.

This Committee, known as the North and North-East Border Defence Committee, would eventually send its report in two parts, according to Mullik, "The first part consisted of its recommendations regarding Sikkim, Bhutan, NEFA and the Eastern frontier bordering Burma. This part was submitted in April 1951. The second part, submitted in September 1951, contained the recommendations on Ladakh and the frontier regions of Himachal Pradesh, Punjab, Uttar Pradesh and Nepal". In the meantime, Lt Gen Kulwant Singh visited the NEFA, and consequently the decision was taken to take over the administration of Tawang.

On November 16, the Political Officer, known as PO in Sikkim brought to the Ministry's notice, "Adviser Shillong[9] signals information received from Walong that 60 Chinese soldiers have captured Drowa Gompa slightly north-east of Rima along with two Tibetan officials who had gone there from Rima". On the border, the situation became more worrisome by the day. The same cable asked the Mission in Lhasa whether the Tibetan Government had any further information about the area.[10] The next day, South Block cabled the Governor of Assam and the PO in Sikkim that a wireless link between Gangtok and Shillong would be operative from November 21. The schedule would daily be from 12.00 to 13.00 hours; Delhi added, "Necessary cipher documents are being sent today and till their introduction cipher messages may be exchanged in Home Ministry's Civil Cipher".

It meant that though the Prime Minister and his ambassador in China did not believe in the Chinese threat, things had started moving at a rapid pace on the ground, owing to the diligence of the Governor of Assam and the PO in Sikkim. Three days later, Nehru admitted to his ambassador in Beijing that he cannot agree with Government's fears and apprehensions about the Chinese. It is interesting to note Nehru's remark in connection with the border with Tibet. The Prime Minister mentioned that, "We realise importance of McMohan [sic] line by which we are going to stand anyhow. We are taking necessary steps on border". Spelling mistake notwithstanding, the Prime Minister realised the importance of the border, but he forgot that the line was born in Simla in 1914 at a time when India recognised Tibet as a separate nation, entitled to sign treaties on its own. In the same communication, Nehru repeated, "Our present policy is primarily based on avoidance of world war".

On November 21, in a strongly worded letter, Harishwar Dayal brought to the notice of the Ministry in Delhi, the implications of the invasion of Tibet for India. He wrote, "the occupation by China of the whole of Tibet or of portions of Tibetan territory bordering India, Nepal and Bhutan, or establishment of a Chinese-inspired regime at Lhasa, will create a variety of problems which are engaging Government of India's attention".[11] But Delhi (read Nehru) was mainly bothered about the role that India could play on the international scene. The Political Officer, one of the brightest young ICS officers, understood perfectly the importance of Tibet for India's borders. He told South Block, "the Chinese have a claim to suzerainty over Nepal and disputes with Burma over the northern frontier of Burma. India's own frontier with Burma in this area is in part not demarcated and in part undefined. We must therefore be prepared for the aggressive assertion of a variety of claims, either directly by the Chinese or through Chinese-sponsored Governments in Tibet and elsewhere".

The PO then gave the historical context of India's border with Tibet, "border disputes and intrigue necessitated negotiation of Anglo-Chinese Convention of 1890 and fixation of McMohan [sic] line as northern boundary of Assam Tribal Areas in 1914. When they occupied Lhasa in 1910 Chinese simultaneously exerted pressure on Bhutan; this move was defeated by the staunchness of the Maharaja of Bhutan and by negotiation of the 1910 Treaty between Great Britain and Bhutan which virtually gave Government of India control over Bhutan's external relations".

Equally, the future ambassador to Kathmandu realised that the situation "is further complicated by the present disturbed conditions in Nepal". The Political Officer added, "In view of the manifest dangers of the situation and of the lessons of history it will probably soon be necessary to warn the Chinese against interference in these territories as well". In Delhi, there was of course, no question of "warning" Beijing of anything; as often, the man on the spot had a wider view of the implications of the unveiling events than the mandarins in the corridors of South Block or the Indian embassy in Beijing. But the latter's voices would prevail. In this remarkable communication, Dayal brought also the state of military preparedness "which is also evidently desirable". The PO explained that the military authorities "are NO doubt considering measures for strengthening border establishments both for maintenance of internal security where this is likely to be threatened and for security against external dangers".

During the previous days, the PO had probably the occasion to share his views with the subcommittee of the North and North-East Border Defence Committee, about the necessity to have a proper road network. Dayal dared to quote a letter from Hugh Richardson sent on June 15, 1949, to the foreign ministry in Delhi. The Scotsman had then suggested that India might consider occupying Chumbi Valley up to Phari "in an extreme emergency". What prompted Dayal to write this letter was probably his meeting with some of the members of the Himmatsinghji Committee, who may have asked him to put his views in writing in order to put some pressure on the pacifists in South Block which could only see the "wider perspectives". The Prime Minister had written two days earlier, on November 19, "I am a little tired of reading the telegrams that come to us from our Mission in Lhasa and our Representative in Sikkim". A remarkably two opposite views of the situation.

## The Rising Chinese Pressure

For a few days, there was a lull in the correspondence. On December 4, Jairamdas Daulatram (written "Doulatram" in the telegrams) cabled Nehru, "Just received following secret cypher telegram dated 3rd from Commandant Assam Rifles SADIYA based on message received from Officer Commanding Assam Rifles WALONG". The Governor of Assam listed the "latest most reliable" information about Chinese communist activity in Tibet, "Chinese communists have their Headquarters at Menkong [located between Rima and Batang] and after capturing Chamdo took into their custody Governor of Dowala [Ngabo Ngawang Jigme] and Shola Lama of Chamdo. Weapons, four three inch mortars, Lmg [light machine guns], Sten, Rifles and pistols, night raids and looting with help of local Lamas". The report detailed the Chinese advance. It was clear that an action was required to defend India's borders.

## The Report of the Himmatsinghji Committee

Even though the records of the Himmatsinghji Committee are absent, it was clear that by the beginning of December, the operation to Tawang started shaping up. The first step taken by the Himmatsinghji Committee was to depute a small Committee of military experts to visit the NEFA and make some preliminary proposals for the redisposition of the Assam Rifles units near the McMahon line. On November 12, the Army Headquarters mentioned that the Committee had been constituted "to examine the possibility of Chinese communist troops occupying the areas in the disputed territory south of the McMahon Line and to consider the feasibility of advancing the outpost of the Assam Rifles in order to forestall such ingress".

The "North and North-East Border Defence Committee" had also discussed the advisability of establishing a post in Tawang. The matter was subsequently referred to the Government "for instructions". Sonia Shukla, who researched the subject, wrote: "while awaiting the government's response, it was decided to establish a small post at Senge Dzong, south of Se La Pass on the road that winds up from Dirang Dzong. But the Indian government was clearly ready to move".[12] Eventually, the reply came from the Government. The decision was communicated to the Adviser to the Governor of Assam[13] on December 10. Rustomji is instructed to establish a post at Tawang "without delay". According to Shukla, "New Delhi's reason for hastening the move of administration into Tawang was the fact that the Tibetans have in the past disputed this territory is, indeed, a very strong argument why we should effectively occupy it before the Tibetans or the Chinese assert their claim to it".

The Government's reply stated that, "in the past we offered an adjustment of the boundary in this area but this can only be done by negotiation and then we could perhaps use it as a bargaining counter for some adjustment in our favour elsewhere. We cannot agree to a unilateral occupation of Tawang by the Communists".[14] Shukla described thus the situation in Tawang, "Already burdened under Tibetan rule, the Monpa tribals were also suffering the after-effects of the massive 15 August 1950 earthquake that had leveled many houses in the region. ... Local accounts say that the Tibetan administration was harsh. They taxed the locals according to their land holdings, made them work for free and took away their agricultural produce. If someone did not give taxes, the Tibetan administration would jail them or take off their shirts and beat them with a cane". This would facilitate Khathing's task.

## Taking the Side of India

As the tension mounted on the remote border, the Adviser to the Governor of Assam knew that he had to win the support of the tribal population to India's side. It was also crucial that the tribal population should not be taken away by the Chinese propaganda. Two days later, on December 13, 1950, Daulatram cabled KPS Menon, the Foreign Secretary; he referred to a note dating August 8, 1949, regarding scale of rations for Assam Rifles. He wanted the paramilitary force to be on par with the Army as far as the rations are concerned. One should not forget that the Assam Rifles function under the Ministry of External Affairs. Daulatram thus submitted, "In view of decision to establish forward defended bases interior hills it is now essential that Assam Rifles should be authorised to draw tinned rations as substitutes for fresh rations. Impossible to procure fresh supplies in the interior hills and equally, it is impracticable to send up perishable supplies from a distance of 9 or 10 days march. In view of developments necessitating dispatch of Assam Rifles in greater strength to distant outposts in hills it is now absolutely repeat absolutely essential that tinned rations should be sanctioned".

The Governor cited the preliminary report of the Himmatsinghji Committee, noting that in its recommendations, the Committee has supported the proposal for the equal rations for the Assam Rifles and the Army. The correspondence available continued to show that the preparations for the Tawang operations were going on.

On December 21, Daulatram informed Delhi about another development, a ciphered wireless message received from Walong says that, "whole Eastern Tibet including Rima now under Chinese. New Governor, Ngabo Ngawang Jigme learnt arrested at Chamdo but his predecessor reported escaped. Whole of Tibetan forces in Menkong Dzokhang Dzong Gartok have been captured". The situation was really alarming with the Chinese coming closer and closer to the Indian border. The message from Walong said, "The headquarters of the troops is located in Trana. Each soldier armed with rifle and a revolver. Vitamin Tablets provided when short of rations".

The message further mentioned that the Chinese troops attack "at night from places least expected and through hills". The message from Walong ended by mentioning that there are "strong rumours that Chinese are picking up our wireless messages".

While Mishmi tribes were claimed by the Chinese "as their people", the Mishmis expected "abundant supply of salt, etc., from Chinese. Mishmis are going up and down and letting out the information off our position". Two days later, on December 23, the Governor Assam sent a telegram to SN Haksar, a Joint Secretary in the Ministry of External Affairs, who is also a member of the Himmatsinghji Committee. Daulatram referred to a message sent on December 16 and addressed to Rustomji regarding Tawang.

Daulatram described the situation thus, "Establishing of outpost possible under present climatic conditions only when winter snow-clothing for Assam Rifles and 200 (the word missing was 'persons' become available). Understand winter clothing for Assam Rifles and Army already ready for dispatch from Kanpur. Please arrange immediate air lift winter snow clothing for 200 persons from Kanpur to Tezpur from where consignment will be taken delivery by Assam Rifles".

Delhi was further informed that "All other arrangements for moving one platoon of Assam Rifles from Dirang Dzong to Towang [it should be 'Tawang', which means 'the blessing of the horse' ('wang' is initiation or blessing, while 'ta' is horse)]. Areas are in hand and move will take place immediately winter clothes made available". According to the estimates of the Governor, it would take six days carrying clothing by porters from Assam plains to Dirang Dzong. He consequently requested that "most essential consignment should be air lifted from Kanpur to Tezpur immediately to avoid delay".

The Governor added a political warning, "Occupation of Tawang may cause resentment in Lhasa but presume action will be taken at proper time to make Lhasa see our point of view".

It would be done in due time by Sumul Sinha, the Head of the Indian Mission in Lhasa. A week later, on December 30, 1950, Nehru instructed the Governor of Assam about three Tibetan officials who arrived from Rima, "We had told you to allow them to come but to watch them. There was a possibility even of their being Chinese spies". The Prime Minister added, "You can deal with them as you like. But I would not like them to be in Kalimpong. Kalimpong today is one of our biggest spy centres". Nehru briefed the Governor, "While Chinese troops have undoubtedly spread out in Eastern Tibet, they have not marched towards Lhasa yet. I do not think there is much chance of their trying to cross our border," though he added, "But we should of course keep full watch there".

## The Takeover of Tawang

India needed someone to not repeat the blunder of Kashmir where large parts of Indian territories were literally "offered" to a hostile neighbour. Some officers were determined to safeguard India's interests. Tawang found its own "Patel" in Jairamdas

Daulatram, the Governor of Assam. In early January 1951, Daulatram ordered a young Naga officer to go and set up the Government of India's administration in Tawang area (then known as Kameng Frontier Division). Till the end of 1950, the entire area from Tawang to Dirang Dzong, south of the Se La Pass was still under some vague Tibetan administration, with the Tibetan *dzongpon* of Tsona in Tibet, collecting "monastic" taxes from time to time in and around Tawang. Maj Bob Khathing[15] of the Assam Rifles then entered the scene.

In 1942, Khathing joined the newly raised Assam Regiment in Shillong and became a captain. He was later advised by Sir Akbar Hydari, the first Governor of Assam after Independence, to join the Assam Rifles. He served with the second Assam Rifles in Sadiya and by 1951 he is inducted in what became the Indian Frontier Administrative Service as an Assistant Political Officer (APO).

An article written by Yambem Laba in the *Imphal Free Press* gave some details, "Summoned by then Assam governor Jairamdas Daulatram (Khathing) was asked, "Do you know Tawang?" He is then given a 'secret' file to study and told to "go and bring Tawang under Indian administration". This task could not be implemented by the British for 50-odd years".

Laba's described the exploit of the Naga officer, Major Bob Khathing who headed the operation, "On January 17, 1951, Khathing, accompanied by Captain Hem Bahadur Limbu of 5th Assam Rifles and 200 troops and Captain Modiero of the Army Medical Corps left Lokra for the foothills, bound for Tawang. They were later joined by a 600-strong team of porters. On January 19, they reached Sisiri and were joined by Major TC Allen, the last British Political Officer of the North East Frontier Agency. Five days later the party reached Dirang Dzong, the last Tibetan administrative headquarters, and was met by Katuk Lama, Assistant Tibetan Agent, and the Goanburras (The Headmen) of Dirang. On January 26, Major Khathing hoisted the Indian flag and a *barakhana* (banquet) followed. The party stayed in Dirang for four days, during which time they received airdrops".

As Laba's article has a few in exactitudes, we therefore need to refer to a biography of Khathing written by Lt Col Bhuban Sing and entitled *Major Bob Khathing, The profile of a Nationalist Manipuri Naga.*[16] Apart from Maj Allen, Khathing's superiors were NK (Nari) Rustomji, the Adviser to the Governor of Assam for the Tribal Areas and SN Haksar, an ICS officer serving as Joint Secretary in the Ministry of External Affairs in New Delhi. One point rarely mentioned was that the local Monpas were delighted by the arrival of the Khathing expedition. The Tibetan 'administration' only consisted in forcefully collecting taxes, which the local people often could not afford to pay; the corvee tax (*ula*) was particularly unpopular.

Interestingly, for years the Chinese Government did not react to the Khathing expedition. A Chinese study on the McMahon line admitted, "Not being clear about the Indo-Tibetan border is clearly reflected in the map drawn by the troops that invaded Tibet". The study further explained, "Regarding the map that the PLA used while invading Tibet, when the 18th Army led by Zhang Guohua invaded Tibet, they still did not have a Tibetan map that they could use. They only had a rough and simple map of Tibet showing subdivisions. There was not even a standard road map. The names of the places and the villages were neither precise nor accurate. This map was found in the archives of the resource committee of the KMT [Kuomintang]; it was made by the British by doing air survey. On the top was inscribed the route followed by Zhao Erfeng while he invaded Tibet (in 1910)".

It is only in 1954 that the Communist regime in Beijing discovered the old KMT maps claiming the entire NEFA as Chinese territory. To return to Laba's narrative, "On February 1, they moved out and halted at Chakpurpu on their way to Sangje [Senge] Dzong". On the third day, they walked a five-mile climb to cross Se La Pass and proceed on to what enters in Khathing's diary as the "Tea Place", where water could be collected from the frozen surface to make tea. By 7.30 hours, the party closed in on Nurunang. On February 4, they reached Jang village where two locals were sent out to collect information and gauge the people's feelings towards their coming. The next day, the headmen and elders of Rho, Changda and the surrounding villages of Jang call on Khathing, "who lost no time in explaining the purpose of his visit and told them in no uncertain terms that they were no longer to take orders from the Tsona Dzongpens. That day, he, Captain Limbu, Subedar Bir Bahadur and Jamadar Udaibir Gurung climbed about half a mile on the Se La Tract to choose the site for the checkpost and construct a barrack".

Khathing's story, as told to Laba, continued, "On February 6 the [Assam Rifles] camped at Gyankar and Tibetan representatives of the Tsona Dzongpons came to meet them. It was also Tibetan New Year or Lhosar, the first day of the Year of the Iron Horse. In the evening it snowed heavily and the villagers took this as a very good omen".

Lt. Col. Bhuban Singh recounted, "On February 6, Bob left for Tawang. The distance from Jang to Tawang was 12 miles. The initial climb of two miles was very steep and this was followed by a gradual climb of four miles up to Sarul ranges. At a bridge across a small stream, before the final climb to Tawang started, the Indian Expedition party was received by representatives of Tsona Dzongpen. The Expedition party camped outside Tawang near Gyankar. The day was the Tibetan New Year Day (First Day of Iron Hare year). In the evening, there was a heavy snow fall and the villagers commented that it was a very good omen".

Laba recounted the story thus, "Next day in the early morning, Khathing accompanied by Captain Limbu and Shri Katuk Lama went to western and then eastern upper slopes which overlooked the ancient Tawang monastery to select a site for the establishment of a permanent administrative headquarters of Assistant Political Officer of Se La Sub-Agency. The selected site should have sufficient area to house a small military cantonment, police lines, civil lines, office accommodation, residential accommodation, schools, hospital and so on, In addition, a parade-cum-playground would also be required, which would consume lot of area. No suitable site was found as the ground was too undulated and broken".

The narration of the historical expedition goes on, "In the afternoon, porters were paid and most of them returned to Dirang area. There was shortage of money too. Some of the porters, who came from Dirang Dzong proper and nearby villages, were told to get payment from Transport Superintendent, Dirang Dzong. With the departure of about 600 (six hundred) porters, the camp looked deserted. The military component of Bob's party was a company of Assam Rifles less one platoon, and therefore had more than 100 (one hundred) men. In addition, the civilian official component was also over 20 (twenty) men. So with arms, ammunition, tentage, ration, camp furniture office equipment documents and stationery etc. the number of porters required was large. Tawang, with just about 300 houses then might have a population of about 2000. The presence of Bob's party of nearly 800 with a substantial number of armed personnel must have been formidable and awesome".

The Assam Rifles of Bob Khathing finally reached Tawang on February 7. The Naga officer later told Laba, "...two days were spent scouting the area for a permanent site where both civil and military lines could be laid out with sufficient area for a playground. A place was chosen north-east of Tawang Monastery and a meeting with Tibetan officials was scheduled for February 9, but they had shown a reluctance to accept Indian authority overnight".

The journalist of the *The Imphal Free Press* who accompanied Khathing on his last trip to Tawang, remembered that the major told him in 1985 that he had no option, but to order Captain Limbu to ask his troops to fix the bayonets and stage a flag march around Tawang "to show he (Khathing) means business". Apparently, it had the desired effect and in the evening the Tibetan officials and elders of the monastery came to meet the Political Officer; they were told that from now on the Tsona Dzongpons or any representatives of the Tibetan Government would no longer exercise any power south of Bumla.

The article continued, "On February 11, Khathing visited the monastery, called on the abbot and presented him and the other monks' gifts that comprised gramophone players, cloth and tiffin-carriers. The next day all the chhgergans (officials) of the 11 tsos (a group of villages) or Tibetan administrative units were called up and a general order was issued directing them not to take any more order from the Dzongpons or Drekhong or pay tribute to them any longer. That afternoon, Tibetan officials and the Nyertsang called for time and permission to exercise their authority till they heard from the Tibetan Government in Lhasa. Khathing put his foot down and told them the "area is ours according to the Treaty of 1914 and there was no question of a reply from their Government in Lhasa and, hence, no extension could be given. Thus, did Tawang effectively become a part of India from that day onwards".

According to Khathing's biography (Bhuban Singh used the Diaries of Khathing for his story) "the morning of the next day, that is, February 8, 1951 was again spent on reconnaissance for site selection, with Captain Limbu in tow. At last, a suitable site was located in the area north-east of Tawang monastery with sufficient area for playground etc. and a good water source. The area was wasteland or khasland, but it seemed to Bob that the NEFA administration had to pay compensation for acquiring the land".

The account of the "takeover" of Tawang continued, "In the afternoon, Bob got busy on the job for which he had been sent and come. He called the Tibetan and monastery officials for a meeting. Notices were served on the two Dzongpens and other officials. Since intelligence reports indicated that the Tibetan officials did not like the Indian presence and had accordingly warned the local Monpas from co-operating with the Indians. There upon the newly arrived Assistant Political Officer of Se La Sub-Agency decided on a show of strength. He informed Charduar and

Shillong about what was happening and sought clear-cut orders to implement the amalgamation of Tawang area to India, by force, if necessary".

Bhuban Singh commented, "Despite the fact that the local Monpas had close religious and cultural ties with Tibet and despite knowing the fact that Tibetan susceptibility might be wounded, Bob was determined to flex his muscle. A nice high-ground close to Tawang Monastery, the seat of power, was selected for meeting the Dzongpens, elders and local people. Bob marched his troops from campsite to the meeting place. His one hundred riflemen formed a box completely encircling the high-ground, a reminder of pre-Napoleonic battle formations. On Instruction from Bob, Capt Hem Bahadur Limbu ordered 'fix bayonet' to his troops. One hundred click sounds of bayonets coming in unison seemed to say "we are even ready for blood." The shining bayonet blades reflected flickeringly the golden rays of the setting sun in a cloudless afternoon of February 8, 1951 at Tawang. The Dzongpens and officials did not attend the meeting. But they must have been watching the scene from peep-holes of the monastery, and receiving the message, however, the crowd which had gathered, must have realised which camp to side with".

The unofficial biography of the Naga officer narrated, "Exuding supreme confidence and exhibiting rare charm, Bob held court for the crowd which included some elders and leaders as also women and children. He spoke to them through interpreter. He told them that the people should not have any apprehension about any interference on their monastic rituals and functioning. Religious freedom was assured by him now and also for future too on behalf of the new administration. He explained to them that the constitution of the new Republic of India tolerated religious freedom and even Godlessness and irreligiousness. As Indians, they would enjoy the same rights and privileges as enjoyed by, say, a Bengali, or a Bihari, or a Maratha, or a Punjabi. All Indians were equal; he hammered into the brains of the Monpas. It is arguably conjectured here that Bob's Mongoloid features and tribal frankness must had produced electrifying trust in what he said to fellow Mongoloid Indians of Tawang. Had a clever and highly qualified say, a Punjabi APO been sent to Tawang, it is doubtful if he could had been as successful as Bob."

Bhubal Singh's conclusion was, "whether Bob subjugated the people of Tawang or liberated them from serfdom is for the world to decide. But one thing is very clear: Bob did his job. Nari Rustomji, in his own words, said that the Government of India could not have found a fitter man than Bob for this job. The crowd welcomed and cheered Bob's announcements, while the Dzongpens and Tibetan officials sulked. Sure enough, the Dzongpens sent message to Lasha (Lhasa) who in turn complained to India's Consul General in Lasha, and ultimately, the complaint went to the External Affairs Ministry through Gangtok in Sikkim".

Using the notes of Maj Khathing, Col Bhuban Singh remarked, "From Bob's side too, wireless messages after wireless messages were sent to Charduar, Shillong and onward to New Delhi giving details of what he was doing. At the same time, he sought approval of Government of India of the actions he had taken and intended to take. Shillong and New Delhi were aghast with what Bob did. They must have preferred a peaceful, non-violent and Panchsheel type of approach. While Shillong was reduced to a mere post-office forwarding information only, lots of consultations and conferences took place in New Delhi and lots of tea were drunk without any decision. In the meanwhile, Bob was told by Shillong to be patient and understanding and above all sympathetic, as if he had terrorised the local people. He was further instructed not to precipitate a crisis".

Looking differently, Neeru Nanda, an IAS officer posted in Tawang in the 1980s also wrote about Khathing and according to her: "Soon after independence, Major Bob Khathing, a Naga officer of the Indian Frontier Service and the Deputy Commissioner, then known as Assistant Political Officer in Bomdila, marched into Tawang. He was greeted warmly by representatives of the Tawang monastery, the three tsorgens (heads) of Choksum (the three chos (or tsos) and other noted leaders who welcomed him with open arms when he declared the intention of the Indian government to establish a permanent office and headquarters in the area. After watching the working of the office and men for about a month the leaders came to him quietly in a deputation with folded hands and grave faces". Nanda continued her narration, "'Well sahib', they said, 'we have been watching your work and we like it but there is something that makes us very suspicious.' 'What is it?' a startled Major Khathing asked, wondering what had gone amiss. 'Sahib', they said melancholically, 'you do not take anything from us by way of tax, neither do you seem to be proposing to take any. This is causing grave concern to all of us.' The sahib relaxed visibly. 'Is that all?' he said cheerfully and drawing himself to his full height".[17]

Khathing then gave a lecture on how there was only one country and India's Government was not exploitative, "The Indian Government considered itself specially bound to develop the brothers and sisters of border areas".

## *The Romantic View*

In this context, the remarks of Nari Rustomji who was responsible to the border areas after Independence are interesting. As he was leaving his job in Shillong to be transferred to Sikkim,[18] he recalled, "I had inherited, in NEFA, an administration that was virtually no administration. I had functioned practically single-handed, issuing instructions by word of mouth during my tours and giving guidance by wireless to officers in our remoter outposts".

But the times are fast changing. Rustomji had to admit that the Chinese invasion of Tibet changed all the stakes, "The Chinese entry into Tibet in 1950 had changed all this political and strategic pressures dictated henceforward a more elaborate and complex administrative structure in the tribal areas than I had favoured, as well as a heavier physical presence of the bureaucracy and of the engines of law and order in the very centre of the hills, and not merely at their extreme southern periphery as in the past. Mine had been governance with a light touch, on a personal, paternalistic basis, in the nature of a man-to-man dialogue with tribal elders. The new policy warranted a difference of technique and it was in the fitness of things that, at this crucial transitional stage, this difference should be symbolically reflected in a change of incumbent at the helm". Retrospectively, it was a grace for India that Governor Jairamdas Daulatram and Maj Khathing acted decisively in Tawang.

On February 9, he wrote to the Foreign Secretary, "The Committee appointed recently to tour the North-East Frontier etc., has taken some action which I consider of doubtful value. For instance they decided to send some troops to Bhutan. I considered this most undesirable and the matter has been dropped".

The Prime Minister commented on the takeover, "Tawang has now been occupied. Probably the step taken was justified. But it was an important step and I should have been consulted about it. The instructions issued to the Officer Commanding should also have been placed before me before issue. This is a frontier matter involving possibly some complications and no step should be taken without full consultation". He later realised that the decisive action of Khathing and Daulatram was the best for India.

## Conclusion

Looking at Maj Khathing's expedition nearly 70 years after the events, one realises that it was the decision of a visionary (Jairamdas Daulatram) and the execution was done by a great soldier and diplomat (Khathing). The area around Tawang had become *de lego* India's territory after the signature on the agreement reached between Sir Henry McMahon and Lochen Shatra, the Tibetan Prime Minister in March 1914 in Simla.

With the advance of the Indian administration towards India's border in February 1951, the area became de facto Indian territory. This was achieved before the Chinese could reach the Indian frontier; in the early 1950s, when they started to reach the areas North of the McMahon Line, India was already in complete administrative control of the entire Kameng Frontier Division. It has to been noted that not a single shot was fired during the military operation and the administration of the Assistant Political Officer (Major Khathing) was welcomed by the local Monpa populations, who for decades had been harassed by the Tibetan officials from Tsona. Retrospectively, the 1951 expedition could be called the "Liberation of Tawang".

### Endnotes

1. National Archives of India, Miscellaneous Paper, File No. 11/30 (Letters, 115 pages, Sardar Patel, Digitized Private Papers).
2. *Selected Works of Jawaharlal Nehru*, Series II (New Delhi, Jawaharlal Nehru Memorial Fund, Volume 15 Part II, page 347); see also *Nehru Papers*, Nehru Memorial Museum & Library, New Delhi, November 20, 1950.
3. Sadiya is located on the banks the Brahmaputra in Assam. It is considered the widest point of any river in the world. In Sadiya, three rivers the Dihang (Yarlung Tsangpo, then Siang), the Dibang, and the Lohit join to become the Brahmaputra. Sadiya is today in Lakhimpur district.

4. Walong is a small cantonment and administrative town in today's Anjaw District of Arunachal Pradesh. Walong is also the easternmost city in India. Anjaw was carved out of Lohit District in 2004. Walong is located on the West bank of the Lohit River, approximately 40 kilometres south of the Tibetan border. Just north of the McMahon Line lies the Tibetan trading town of Rima. Incidentally, it was the epicentre of the August 1950 Earthquake.
5. Drowa Gompa is a place located in Tibet, north of Anini in today's in Dibang Valley district of Arunachal Pradesh.
6. The north-west of Rima lies North of Kibithoo which is the last Indian post in Anjaw district of Arunachal Pradesh.
7. As per the reply given to an RTI petition. See, https://www.timesofindia.indiatimes.com/india/MoD-cant-locate-five-key-reports-on-military-reforms/articleshow/10347823.cms.
8. Mullik, B.N. 1971. *My Years with Nehru—The Chinese Betrayal.* Delhi: Allied Publishers.
9. Nari Rustomji, Adviser to the Governor of Assam for Tribal Areas (NEFA).
10. Kongpo of Southern Tibet.
11. Letter available in the Nehru Papers (JN Collection) held at the Nehru *Memorial* Museum and Library in Delhi. The "papers" are not indexed (November 21, 1950).
12. Sonia Shukla, *Forging new frontiers: Integrating Tawang with India, 1951*, (China Report, November 2012, 48(4):407–426).
13. Chakravarty 1953, DO no. D. 232-NGO-50 in CGA 140/50, 146.
14. Ibid.
15. Born on February 28, 1912, in Ukhrul district of today's Manipur, Ranenglao (Bob) Khathing belonged to the Tangkhul Naga tribe.
16. Lt Col H Bhuban Singh, *The profile of a Nationalist Manipuri Naga*, Pritam Haoban publisher, Imphal, 1992.
17. Neeru Nanda, *The Land of Mon*, Vikas Publishing House, 1982.
18. Rustomji was recalled to Shillong in 1959 for a second tenure as Adviser for the frontier areas.

**Claude Arpi** is an Independent Researcher on Tibetan Studies based in India. A French national (from Angouleme) settled down in South India since 1974. He is the author of several books and more than one thousand articles on Tibet, China, India, defence and border issues and Indo-French relations. He holds the Field Marshal K.M. Cariappa Chair of Excellence of the United Service Institution of India for his research on India-Tibet Relations. His website (http://www.claudearpi.net) carries the largest number of historical documents on topics such as the Indo-China relations, the flight of the Dalai Lama in 1959 and the consequences of the 1962 Sino-Indian war for India, etc.

# Chapter 5
# Monpas, Tawang Monastery and Tibetan Buddhism: Ethno-Religious Links

**Tsetan Namgyal**

## Introduction

The etymological meaning of Tawang (*rta dbang*) is such though it is made up of two terms ta or *rta* meaning 'horse' and wang, or *dbang* means 'authority' or empowerment of something very auspicious and superiority.[1] It symbolizes the place with many horses where most of the works are done through the horses during those times in this particular part of the world called Tawang. In traditional Tibetan Buddhism, especially Tibetan Tantric Buddhism, horse is a significant ritual entity in practice. However, in the context of understanding the place of Tibetan Buddhism for the 'Mon', the term 'Mon' has a different meaning and application as per Tibetan tradition.

The literal translation of the Tibetan term '*Mun*' is dark, darkness or ignorance. The symbolism was drawn from the perception that reiterated the same as '*Monpas*' were not followers of Buddhism; hence, they are '*Mun*' as far as their mind and wisdom are concerned. Thus, later many scholars and sources by mentioning '*Monpas*' meant unenlightened, ignorant, uncivilized or even barbaric as was the descriptions formed against the '*Monpas*' by the Buddhist Tibetans who were not following Buddhism. In the Tibetan Buddhism, monastery controls everything from socio-religious to the political economy, all under the garb of the one single institution a kind of religious symbolic entity, i.e. monastic institution. In a sense, monasteries are not just a representation of the religious activities but have been a symbol of political identity ever since the traditional Tibetan Buddhism has become the most significant identity culture of Tibet and Tibeto-Himalayan regions.

It is been proclaimed that the *Monpas* of *Monyul,* known as the land of Mon, from 500 BCE to 600 CE, was once under the sovereign rule of the *Monpas* but later reduced as the region came under the sporadic control and power of Tibet and it

T. Namgyal (✉)
Centre for Inner Asian Studies, Jawaharlal Nehru University, New Delhi, India
e-mail: tsetannjnu@gmail.com



M. Mayilvaganan et al. (eds.), *Tawang, Monpas and Tibetan Buddhism in Transition*,
https://doi.org/10.1007/978-981-15-4346-3_5

became the periphery to Tibet that changed the demographic map of Monyul in due course of time. There are numerous Tibetan literatures and other sources on '*Monyul*' and '*Monpas*' in both secular and religious texts that have mentioned about this land of the Mon and people.

*Monpas* are in general an ethnic community that primarily extended to Tibetanized trait in language and culture that spread across the whole of Himalayas as a separate geographical entity from eastern kingdom of Bhutan, Sikkim, Ladakh (Jammu & Kashmir), Baltistan (Pakistan) to Lahaul and Spiti (Himachal Pradesh). In contemporary time, most of the Monpas are still residing in the main southern part of the landmass of India's Himalayas, which traditionally called 'Lho Mon' (southern Mon or inhabitants of southern part of the Himalayas). Incidentally, this region is claimed by the Chinese today as Southern Tibet.[2] On the contrary, there is also called 'Shar Mon' the 'Eastern Mon' that referred to Bhutanese 'Mon' in particular though it includes the Eastern Himalayas region of Sikkim that comprises of the Monpa ethnic groups of people.

## *Contour of Tibetan Culture Area*

In the context of the Tibetan culture, most important facet will be the historical connectivity and linkages between the concerned land and people, often corroborated by recorded history. But of the many contributions, the role of the Tibetan monks in strengthening the historical relationship and linkages between Tibet and its other parts of greater cultural area like Tawang is notable. A persona like the sixth Dalai Lama Tsangyang Gyatso, an ethnic Monpa, who happened to be born in Tawang in Arunachal Pradesh, India, 'Lama Zhabdrung Nawang Namgyal' from Bhutan, 'Arhat Bakula' from Ladakh, India, and 'Agvan Dorjeiv' from Kalmyk, India, all have played an important role in the development and expansion of Tibetan Buddhism through socio-religious and political activism.

In the words of Gephart and Waldenfels (1999), the traditional Tibetan Buddhism should be understood as an institutional web of monastic institutions covering all the areas inhabited by the Tibetans, and it denotes in another way the entire Tibetan cultural area from Trans-Siberia to Trans-Himalaya to Baltistan. Therefore, ever since the Tibetan Buddhism reached or was introduced in these regions of the Himalayas, it has become an integral part of the social fabric and identity of the inhabitants in the region.

## Monpas and Tibetan Buddhism

Traditionally, Arunachal Pradesh is known for its different stocks of colourful indigenous tribes, where there are as many as twenty-six very significant progenies with their different prisms of history and cultural heritage residing for ages. One can say that immigration from all the parts mainly from Tibet southern-west and other northern and western parts of India even Myanmar had made Arunachal Pradesh the ethnic mosaic of indigenous social folk culture. However, they are distinct from each other, due to their separate geographical sphere of social and religious boundaries within one state. Each of them has their certain distinctive identity culture and such characteristic that reflect in their language, dress and food customs, and culture.[3] As far as language is concerned, there are more than fifty distinct languages and dialects in addition to many sub-dialects spoken exclusively across the state.

Arunachal Pradesh is known for its diverse ethnic and tribal groups. There are as many as twenty-two significant tribes and numerous sub-tribes, which are said to be of Sino-Tibetan origin.[4] Presently, all these tribes have been recognized as Scheduled Tribes under the constitution of India.[5]

Within these tribes, the Monpa tribe belongs to the region of Mon (Monyul[6]). Ancient Tibetan historians usually referred Mon and Monyul to the border areas of Southern Tibet, including today's Arunachal Pradesh. In addition, there are few other traditional Tibetan literature such as '*rnam thar*' (biography) and '*chos byung*' (religious history) written by different Tibetan scholars during the second phase of diffusion of Buddhism (bstan ba phyi dar) have also found mentioning such places as 'Lho Mon gyi yul' which means the country of Southern Mon. People settled around today's Tawang and West Kameng, and who follow the traditional Tibetan Buddhism is identified as Monpa or the country of Mon.[7] The Tibetanization of Monpas with Tibetan culture and language took place like the other Himalayan Tibetan ethnic groups such as Lepchas, Bhutia, Tamang, Lamas and Ladakh. As far the pre-Buddhist religion in the region of Tawang is concerned, the religion before the arrival of Buddhism was the Bon religion. Tawang came in the eleventh century under the influence of the Buddhism.

Besides Buddhism, most in Arunachal Pradesh profess their indigenous religion called Donyi-Polo, an animistic form of nature worship along with Christianity and Hinduism. Within Buddhism, interestingly there are some tribes, who follow the Theravada Buddhism, such as Khamtis, Singphos and TikhakTangas, while Monpa, Sherdukpen, Nah, Memba, Khamba, Meyor and Zhakring are very firm adherent of traditional Tibetan Buddhism, who also reside along the Indo-Tibetan border area of the state.

The different stocks of people practising their own indigenous culture are the core segment of Arunachal Pradesh in general and Siang and Tawang area in particular where some different aboriginals are found living harmoniously with Mon, Brokpas, Bjobs Sherdukpen, Khamba, Memba, Nah, Meyor, Zhakring (Tibetan stock) and other Indo-Tibetan, Indo-Malayan and Indo-Aryan groups.

The Monpas are settled along the entire Tawang and West Kameng regions such as Dirang, Bomdila and Khalaktang Sherchokpa, Rupa, Shergaon and Sherdukpen. The Tibetan Buddhist epic '*Gesar Ling*' has also found mentioning numerous occasion the term 'Mon' as a hegemonic people of pre-Buddhist Bhutan besides *Khen, Brokpas, Birmis*, and Koch, etc. It also found mentioning in various Bon literatures and folklore narratives such as a monumental work of 'Juthing' by Ju Mipham Jamyang Namgyal (1846–1912) where he has mentioned the land of Mon.[8] In this context, *Zhangzhung*[9] referred to the Western and Northern belt of ancient Tibet once a separate independent kingdom where it also found referring to the country of 'Mon' as 'Mon' might possibly be the corrupt word of 'Bon' because the traditional Tibetan sources also claimed that 'Bon' was originated in *Zhangzhung*. In the later period, it might have flourished into the entire Tibetan cultural areas after the ages of conflict and clashes with Buddhism during its interception period in Tibet. The first passage describes the Unicorn 'Dralab Se'uRugchig' as one of the important manifestations between Drala and Werna in Bon religion. In a way, it is a representation of the dimension of existence of the *Ye rjesmonpa'i sgrabla* also known as *smon pa*.

The other important historical religious significance of Tawang is that the sixth Dalai Lama (1683–1706) *Tsangyang Gyatso*[10] who was born in Tawang then was part of the southern borderland of Tibet. Hence, ever since Tawang becomes a sensitive place both religiously and politically, China's claim over Arunachal Pradesh has drawn more from the religious significance that Tawang holds in being the birthplace of the sixth Dalai Lama. Also, China is aware of the historical connection and the nature of the smaller states like Bhutan, Sikkim and Tawang in being protectorates of Tibet, thus indirectly linked to China's historical narrative of the Southern Tibet.

Tawang and Monpas were historically and traditionally under the authority of Lhasa, as a part of Southern Tibet.[11]

## Tawang Monastery—Ethno-Religious and Sociopolitical Edifice

Tibetan chronicle text called '*gro ba bzangmo'irnamthar*' has mentioned that a king named '*Ka la dbangpo'i pho brang*' (monastery) consecrated and situated that is also called '*rta nag Mandal*' is the present Tawang Monastery.[12] However, oral and historical record says Tawang Monastery is a seventeenth-century legendary Buddhist vihara affiliated to '*Gelug pa*' school of traditional Tibetan Buddhism and also one of the largest and oldest existing Tibetan Buddhist monasteries in Asia.[13] It also known as '*Yul Mandal Gang*' as it is said that Tawang looks like a 'Mandala' from the place where the monastery is situated.[14] In true sense, the entire Tawang including the Tawang Monastery was an integral part of the Tibetan province called 'Tsona', and it remained under the Tibetan jurisdiction till 1951. It was the winter place of the *Tsona's* district head (*Dzongpon*), while *Tsona Dzong* was the summer place, which is now in Tibet near *Chumbi* Valley around Lhasa and *Shigatse*.[15]

As Capt. Pemberton, a British Indian member, in his 1838 report on Bhutan referred the region as a tract of country dependent on Lhasa and forming an integral portion of Tibetan territory.[16]

*Source*: https://www.google.co.in/search?q=tawang+monastery+photos

There are mainly two sources based on Tibetan tradition concerning the Tawang Monastery. The first sources illustrate what the oral tradition and other circumstantial shreds of evidence acclaimed: the earliest *Gelug pa* history in Tawang begins with the '*Thangtong Gyalpo*', a *Kagyupa* lama from Tibet who lived from 1385 to 1462.[17] He was considered the disciple and associate of the first Dalai Lama, *Gedun Drub* (1391–1475 AD).

The other version is of well-known lama called *Tanton/Thangtong Gyalpo* (1385–1462) better known as a saint engineer and patron of Tibetan opera who built many bridges across rivers in and around central Tibet. He also came to Tawang and meditated in a cave near *Karling* towards the south-east of Tawang. He is credited of building the suspension bridge (*Chaksam*) over Tawang river near *Karling* village south-east of Tawang, and the bridge still exists which connects *Mukto* and *Kitpi* in the Tawang districts.

Tawang Monastery is also called by '*Galden Namgyal Lhatse*' (the Celestial Paradise of the Divine site chosen by horse) and is one of the most prominent Tibetan monasteries in India as its root affiliation with the *Drepung* Monastery in near Lhasa. The Tibetan also used implicitly or explicitly Tawang as '*Loyul*' '*Lo*' means south and '*Yul*' means country therefore, on the whole Tawang is referred to Southern country. After the establishments of a full-fledged religious and political power systems over the whole of Tibet from the fifth Dalai Lama until seventh Dalai Lama, the mandate was also to the whole Tawang area including the Tawang Monastery which was specifically significant for the Dalai Lama's institution because the sixth Dalai Lama was born in *Monyul*, Tawang, in a *Monpa* family. When the present Dalai Lama exiled Tibet in 1959 and crossed over to India on 30 March 1959, he first visited and spent a few days at the Tawang Monastery before moving to *Tezpur* in Assam on 18 April 1959.

Today, the Tawang Monastery is an important seat not only for the followers of 'Gelug pa' school of traditional Tibetan Buddhism in particular but also for Monpas in Tawang. It was founded by Lama Merak Lotos/Lodre Gyatso (Tib. *Me rag bLa ma bLogros rGyamtsho*) in 1680–81 as per the wishes and auspicious direction by the great fifth Dalai Lama (*Nawang Losang Gyatso*)[18] (1617–1682). He was perhaps the most notable person in the political history of Tibet besides being an erudite scholar who has succeeded in controlling the greatest part of Tibetan area, but such power and authority cannot be called absolute control. It says that the lama has abandoned his former monastery called *Galden Tselling Gonpa* in a village called '*Mera*' in eastern Bhutan. He came to Tawang at the request of fifth Dalai Lama and founded the Tawang Monastery. He was well known as Mera lama by the locals. It was when the Mongol king *Gursikhan's* troops invaded Bhutan in 1644.[19] The period was also when the sectarian struggle between the *Drukpa* and the *Gelug pa* schools was at boiling phase. Though in 1646 Tibet and Bhutan reached a peace agreement restoring the status quo, it could not sustain and hostility developed again. It was during this period the Mongol and Tibetan troops came back to assist the *Monpas* forces in 1647.[20]

Notably, the fifth Dalai Lama requested the Mongol prince Gursi Khan to help Tibet in their war with *Drukpas* of Bhutan and accordingly the Mongol troops helped Tibet with invading Bhutan and established the supremacy of Gelug pa against the '*Drukpa Kagyud pa*'.[21] According to some oral tradition of Tibetan Buddhist records, a Lama of Tawang Monastery referred to one Mongol commander '*Sokhpo Jomkhar*' as one who helped the Gelug pa in their struggle with Drukpas of Bhutan.

At the same time, the monastery has witnessed many revolts and events pre- and post-phase of fifth and sixth Dalai Lamas, especially after the demise of the great fifth Dalai Lama. The Mongol *Lhazang Khan* started an invasion to depose the sixth Dalai Lama with the support of *Kangzi* emperor.

Since, it belongs to the *Gelug pa* sect of traditional Tibetan Buddhism, Tawang's religious and political affiliation with all the Gelug pa monasteries of central Tibet such as *Sera, Gaden, Drepung* and *Tashi Lhunpo* remained till the beginning of the Chinese Cultural Revolution. Today, the designated official head of the Tawang Monastery is Dalai Lama though there is an abbot as a head of the monastery who looks after all the monastic activities. This makes Tawang an important symbolic benchmark of Tibetan Buddhist identity in exile.

In fact, the symbolism of Dalai Lama over Tawang is seen by China as the substantial reason for its claim on Arunachal Pradesh as Southern Tibet. Not only Tawang's historical connection with the Tibet's *Sera, Gaden, Drepung* Buddhist monasteries is significant, these historic connections also make it a part of the narrative on the Greater Tibet. During the prosecution of Buddhism in India by different dimensional forces particularly by the Turkish armies who destroyed the world largest Buddhist monastic centres/universities of Nalanda, Vikramshila, Taxila, many Buddhist monks fled and reached Tawang as a refugee. The monks were then given shelter in the Tawang Monastery and other places in the region.[22] Later, they then built new Buddhist temples in and around Tawang and kept preserving all the sacred Buddhist objects including texts whatever they could carry during their escape.

Until 2008, there was no head lama or Rinpoche appointed by *Monpas* or the Tibetan community, but later in the wake of a realization that a vacuum remains in administering the monastery, the Dalai Lama for the first time appointed a local

*Monpa* Lama from Rama village south of 'Se La' as head of the Tawang Monastery in 2008. He was the first Indian monk who was appointed as the head lama of Tawang *dGonpa*. Until the tradition of selection or appointing Tibetans from Lhasa remained enforced and it had been selected and appointed for the highest post from Lhasa by the Dalai Lama.

Along the side of Tawang Monastery, there are three other the most ancient and oldest Buddhist monasteries[23] which are associated with the *Nyingma pa* school of traditional Tibetan Buddhism built by Lama from Bhutan called '*Sherbum*' sometimes around eleventh–twelfth century AD in Tawang. As oral history, it was *Padmasambhava* (*Tib. Slopon Padma 'byungnas*) who once professed the construction of these three monasteries during his visit to this auspicious place of Tawang.

## *Repository of Buddhist Literature and Other Literary Works*

The monastery has an extensive collection of Tibetan Buddhist scriptures, i.e. *Kangyur* and *Tangyur*,[24] in the Xylograph form, and a lot of other classical literary works by Indian and Tibetan masters. Apart from these illustrated manuscripts and literatures, there are a huge collection of murals and Thangkas paintings and other religious artefacts. In addition, there are also some rare Buddhist manuscripts of the textual composition of the five major and the five minor fields of knowledge on different Buddhist doctrinal studies including sutra and other Buddhist hymn texts. Most of the manuscripts were brought from Tibet at a different historical period, especially during the cultural interaction with Tibet and some of them written in the Tawang Monastery by local learnt lamas. Yet, it is said that most of the existing scriptures and texts were brought from *Drepung* Monastery around the seventeenth century during the completion stage of the monastery.

Overall the number of collected rare Buddhist manuscripts including xylograph (woodblock painting) in Tawang Monastery is around five thousand[25] at present. Although, there are more than a dozen different other Buddhist monasteries and mostly associated with *Gelug pa* school of traditional Tibetan Buddhism. Overall, there are as many as more than one hundred Buddhist monasteries in Tawang area alone, which have their extensive collection of manuscripts and other Buddhist texts and xylographs are well preserved in their respective monasteries.

The following are some exclusively most significant collections of the Tawang Monastery as follows:

- Three sets—two sets are handwritten, and one set is in printing blocks—of ***Kangyur* (*bkagyur*)** while the printed set covers to 101 volumes and one handwritten set has 131 volumes and the other 125 volumes.
- Two sets of **Tangyur/*bsTangyur*** (each set contains 225 volumes).
- Five volumes of ***Changla Sungbum***.

***Gyastong pa*** (means hundred thousand verses or parables) is one of the most important Tibetan religious texts which read out at the very special occasion was written in gold and silver. Besides, the Gyastong pa there are few more rare manuscripts which have been brought from Tibet during the construction process of Tawang monastery, however due to non-availability of confirmed records it could not be ascertained as where and when it received from but according to some local lamas it says that these Buddhist texts along with others Buddhist religious objects were brought from Tibet at different period of time by different lamas.

## *Today's Sociopolitical Edifice*

The Monpas are acclaimed as mountain agriculturist because of their generic acclimatizing power to the hard and harsh climate condition and adaptability endurances. Today, the Monpas are in the phase of transition and transformation mainly from post-period of China–Indian border conflict of 1962 as it resulted in securitization of their place with very rapid pace of construction of road, bridges and communication network, etc.

For the Monpas, everything is interconnected and evolved with their local ecosystem and this is one of the most core essences of their sociocultural ethos and value. The religious rites and ritual practices, which are completely blended with their traditionally oriented cultural milieu, the spiritual belief system that they have been following in their day to day life and work while firmly believing and following with ethical and moral values. Today, the socio-economic changes have brought a significant effect on the society and their social custom as a result, their age-old custom rituals and belief system, including the significance of Tawang Monastery that is fast changing under the wave of modernization.

The Tawang Monastery, which uses to play a dominant role in the sociopolitical life of Monpas, is reduced to symbolic today. Monastery and the abbott (head) are kept themselves away from political arena. However, the abbott is consulted by the locals on various important issues concerning the Tawang and Monpas. The monks and lamas who study in the monastery, however, keep their affiliation with the local organizations and political activism.

## Conclusion

The Mon and *Monpas* of the Arunachal Pradesh in particular and rest of the entire Trans-Himalayan *Monpas* are being disseminated and riveted into the plain area of India's Himalayas and lead a different life except some segment of Bhutanese *Monpas* who still live by their profession of hunting and professing nature.

Indeed today the geopolitics of Tawang and Tawang Monastery have become the borne of the contentious issue between India and China in both rims of its strategic and political compass primarily only because of its Tibet centric issue. Moreover, the Tibetans in general and particularly the present Dalai Lama who is residing in India since 1959 are always been under the scanner of watchful Chinese eyes. The Dalai Lama for China is not merely a religious symbol, but he has been a political entity, and China is very sensitive to both the physical movement and the verbal utterances of the Dalai Lama. As an Indian, we all know that Tibet question is a very sensitive matter in the milieu of China relationship with India. And it will remain so until the Tibet and Tibetan issue resolved.

Also, any Buddhist monastery irrespective of its affiliation in the Trans-Himalayan region is important for Tibet and Tibetans not only for those who are residing in India as a refugee but the entire Tibetan diasporas across the world. However, particularly the Tawang *Gompa* is significant because it forms a crucial part of the Tibet's national identity in general, and more it represents the local identity culture of '*Monpa*' ethnic people who constructed and developed it into a world Buddhist heritage under various phases of its political and religious jarring circumstances during the edifice time of its establishment. Hence, today, unfortunately, it has become a centre of political activism of different religio-political staking parties and groups. Today, it's certainly to be noted that the monastery is an important epitome of Buddhist identity and culture of the entire Buddhist world though here it's symbolizing the Tibetan identity and culture in particular and it has been playing its role in shaping the identity politics of Tibetans in exile.

On the other hand, Tawang Monastery is seen by China as the strong reason for its claim on Arunachal Pradesh as Southern Tibet because of its historical connection and linkages with three largest monasteries of Lhasa (*Sera, Drepung and Gaden*) in particular and most importantly considering it as an integrated part of Tibet which is now a part of China.

## Endnotes

1. There is a short story that how the term 'Tawang' came up in the process of its consecration and construction of the monastery by the founder Monpa lama of Tawang called Merak Lama. While he went on in search of a proper auspicious place to build a monastery, his horse (rta) kept getting lost always where he left. And each time he lost his horse, he found it at the place where the present Tawang Monastery is situated. He then thought to himself that the place to be the right and auspicious and immediately decided to consecrate the monastery. From then, the place is called Tawang, which means horse empowerment or authority.
2. In that sense, the region of Tawang and the Tawang Monastery has been become a contested issue between India and China particularly after the Chinese taken over Tibet in the 1950s.
3. For details, see, Government of Arunachal Pradesh, http:arunachalpradesh.nic.in/peole.htm.

4. Osik (1986), p. 4.
5. Jha (1985), p. 68.
6. Traditionally the term 'Monyul' that usually refers to the Tsona district of ancient Tibet and the entire Tawang tract along the Himalayan range.
7. Ibid, p. 41.
8. Norbu (1995).
9. Sahngshung is situated in the western part of Tibet, and according to Bon religion, the seeds of wisdom and spirituality have originated from this place of ancient Tibetan kingdom, which later by western scholars validated that the Bonpos teaching was imported from Tag zig to Zhangzhung and Tag zig has identified a region with Persia the present Iran where Buddhism was flourished through the caravans through the crossroad of silk route.
10. He was born in the present Indian state a district city called Tawang of Arunachal Pradesh in the north-east frontier of India's Himalayas. There are lot of ambiguous sayings about the sixth DL, and also that much of vague documents found stating that in the seventh century a local chieftain was sent presently to the king of Tibet. He was the most controversial and intriguing characters in the entire history of Tibet in both the political and religious realms in Tibet. As soon he took the throne as a ruler of Tibet in 1701, sooner he was deposed by the Mongol Dzungar leader Lozang Gyatso in 1706 and followed by the imperial commander deported in 1720 to China where he died in 1725 in very mysterious circumstances. Refer to Petech (1972).
11. It is only in 1914 during the reign of the 13th Dalai Lama that Tawang was granted to India under the British Raj at the Simla Convention. At this point, an agreement was also made over on the location of the McMahon Line, which was agreed upon as the boundary between British India and Tibet.
12. Ibid, p. 41.
13. Namgyal (2014).
14. Ibid.
15. Nyman (1976).
16. ibid.
17. Snellgrove and Richardson (1968), p. 42.
18. He was a great scholar who knew Sanskrit besides having all kinds of skills and leadership qualities; hence, he was the chief architect in bringing peace and unity among all segments of Tibetans for a united Tibet. He also built the palace of Potala in 1645.
19. Rahul (1971), pp. 23–24.
20. Since the Mongols and Manchus used Dalai Lama for their political gains, on the contrary, the Dalai Lama has also utilized the Mongols and the Manchus for the realization of his own political and religious agenda. The fifth Dalai Lama's influence was so prominent in Tibet's politics besides being one of the great scholars who knew Sanskrit and Indian culture and philosophy.
21. Rahul (1971), pp. 23–24.
22. Sharma (1993), p. 91.

23. These are Urgyelling, Sangelling and Tsorgelling; all are situated near the proper city of today's Tawang.
24. Tibetan canon is called Kangyur and Tangyur two separate vast of literary works initially composed based on Buddhist Sanskrit canons of 'Tripitaka' which translated into Tibetan and later disposed into two broad divisions called Kangyur and Tangyur.
25. http://ir.inflibnet.ac.in:8080/ir/ViewerJS/#../bitstream/1944/1/16.pdf.

## References

Aris, M. 1979. *Bhutan: The Early History of a Himalayan Kingdom*. Warminster: Aris and Philips Ltd.

Beal, S. 1996. *Buddhist Records of Western World*, vol. II. London: Kegan Paul.

Bod kyi lo rgyus rig gnas dpyad gzhi'i rgyu chhab dbsab sgraigs/ bod rang skyon gljong schhab grogs rig gnas dpyad gzhi'i rgyu chha au yon lhankhang gi srtsyom sgrig byas pa// published by Mi rigs dpyes krunkhang, Dharamsala, HP India.

Chakravati, B. 2003. *A Cultural History of Bhutan*, vol. I. Kolkata: Sagnik Books.

Dai, Manang. 2009. *Arunachal Pradesh the Hidden Land*. USA: Penguin Enterprise group.

Das, S.C. 2007. *A Journey to Lhasa and Central Tibet*. New Delhi, India: Rupa & co.

Dhar, Bibhash. 2005. *Arunachal Pradesh the Monpas of Tawang in Transition*. Guwahati, Assam India: Geophil Publishing House.

Dutta, S. 2002. *Cross-Border Trade of North East India: The Arunachal Perspective*. Delhi: Hope India Publication.

Francke. 1975. *A History of Ladakh*. New Delhi: Sterling Publication.

Jha, D. 1985. *The Wealth of Arunachal Pradesh*. Delhi: Mittal Publications.

Kumar, B.B. 2002. *Border Trade in Arunachal Pradesh: An Historical perspective*. Delhi: Hope India Publication.

Lamb, Alastair. 1966. *The McMahon Line*, vol. I, II. London: Routledge & Kegan Paul.

Metha, P. 1980. *The North Eastern Frontier: A Documentary Study of the Internecine rivalry Between India, China, and Tibet*, vol. 2., 1914–54. Oxford University Press.

Namgyal, Lhamu. 2014. *Stories from Monyul*. Dharamsala, HP: Library of Tibetan Works & Archives.

Norbu, N.Drung. 1995. *Deu and Bon (Narrations, Symbolic Languages and the Bon Tradition in Ancient Tibet)*, 58–59. Dharamshala, HP, India: Library of Tibetan Works and Archives.

Nyman, L. 1976. Tawang-A Case Study of British Frontier Policy in the Himalayas. *Journal of Asian History* 10 (2): 151–171.

Osik, N.N. 1986. *A Brief History of Arunachal Pradesh*. New Delhi: Omsons Publication.

Petech, L. 1972. *China and Tibet in the Early eighteenth Century*, 17–20. Leiden.

Phuntsho K. 2013. *The History of Bhutan*. Random House India.

Rahul, R. 1971. *Modern Bhutan*. New Delhi.

Ramble, C. and M. Brauen. 2008. *Anthropology of Tibet and the Himalaya*. Kathmandu, Nepal: Vajra Publications.

Rizvi, J. 1983. *Ladakh: Crossroads of High Asia*. YMCA, New Delhi: Oxford University Press.

Rta wang dgonpa'i lo rgyus Mon yul gsal ba'i me long/ rtom pa po or composed by rgyal sras sprtayulsku, Published by Amnye Machen Institute, Dharamsala(HP) India.

Sarkar, N. 1981. *Tawang Monastery, published by Director of information and Public Relations*. Shillong: Government of Arunachal Pradesh.

Sbas yulskyid mo ljong skyi chos 'byung bzhugs so/ Mon gyi bstan rig gzhung 'zhinskyong tshogs pa, published by Buddhist Culture Preservation society, Bomdi la, Arunachal Pradesh, India, 2002.

Schaeffer, K.S., and M.T. Kapstein. 2013. *Sources of Tibetan Tradition*. New York: Columbia University Press.

Shakapa, T.W.P. 1967. *Tibet: A Political History*. London: Yale University Press.

Sharma, K. 1993. *Painted Scrolls of Asia*. New Delhi, India: Intellectual Publishing House.

Sharma, Ramesh. 1983. *Images of Sikkim: The Land, people, and culture*. Rigsum Publications.

Snellgrove, D. and H. Richardson. 1968. *A Cultural History of Tibet*. London.

Tenpa, L. 2018. *An Early History of the Mon Region (India) and its Relationship with Tibet and Bhutan*. Dharamshala, India: The Library of Tibetan Works & Archives.

Vohra, S. 1993. *The Northern Frontier of India-The Border Dispute with China*. New Delhi, India: Intellectual Publishing House.

**Tsetan Namgyal** is an expert on Buddhist and Tibetan Studies and presently teaching at the Center for Inner Asian Studies (Tibetan and Himalayan studies) in the School of International Studies, Jawaharlal Nehru University (JNU), New Delhi. He also taught at the Rastriya Sanskrit Sansthan (Deemed University Lucknow) and Visva Bharati University, Santiniketan, India before joining JNU in 2009. He has published over thirty scholarly papers in national and international reputed journals including his recent four volumes of books titled, '*Trans Himalayan Tibet, A Copter Approach*.'

# Chapter 6
# Lifestyle and Health Issues among the Monpas of Tawang

**M. Mayilvaganan**

## Introduction

### *Lifestyle and Sociocultural Profile of Monpas*

The Monpas are predominant inhabitants of the westernmost regions of Arunachal Pradesh, chiefly, the districts of Tawang and West Kameng. Within these two districts, their settlements are predominantly in the areas of Zemithang, Tawang, Dirang and Kalaktang. Depending on the place of living, they are often called as 'Tawang Monpas' or 'Northern Monpas' or Dirang or 'Central Monpas', and Kalaktang or 'Southern Monpas'.

The language spoken is though 'Monpa language', it falls under the Bodic group of the Tibeto-Burman language. But within the tribes, the Kalaktang Monpas, Tawang Monpas and Panchen Monpas (Zemithang) use different dialects of the same language for communication. Dirang and Kalaktang Monpas use a dialect of Bhutanese Brokpa language predominantly, whereas Zemithang–Tawang Monpas use a dialect of Tibetan–Bhutanese Dakpa language.[1] In Dakpa language, the term 'Mon' and 'Pa' signify the 'Men of the Lower Country' or the inhabitants of southern regions to Tibet, i.e. Tawang. Nevertheless, many other aspects of their life are quite similar. The Monpa places of settlement are often situated on the slopes of the hills or in the valleys. Some of the villages on the slopes of the hills do not even have proper road and necessary facilities. Tawang and Zemithang Monpas, by and large, live in slopes of the hills except the Monpas in Tawang Town, while Monpas in Dirang and Kalaktang live in the valley.

M. Mayilvaganan (✉)
International Strategic and Security Studies Programme, National Institute of Advanced Studies, Bengaluru, India
e-mail: mayil@nias.res.in



M. Mayilvaganan et al. (eds.), *Tawang, Monpas and Tibetan Buddhism in Transition*,
https://doi.org/10.1007/978-981-15-4346-3_6

The Monpa people are Buddhist tribe belonging to the Lamaistic School of Tibetan Buddhism of the Mahayana sect, namely Gelukpa and Nyngmapa sect. Before adapting to Buddhism, Monpas were practising an indigenous religion called Bon,[2] which existed in Tibet as well. Interestingly, yet a number of Bon deities are worshiped as protectors within the Buddhist pantheon these days. Besides, a small or big 'Gompa' (Buddhist monastery) is noticeable in every Monpa village, often situated on the elevated place in the vicinity. The 'Phan' (prayer flags), 'Mane' (stone shrines) and 'Chorten' (small chapels) decorate the roadside.

The houses of Monpa are usually double- or triple-storied building, made mainly of locally available stone and generally on the surrounding or nearby the Gompa. Each house has the local traditional steam heating system called Bukhari, food storing room and a prayer room (chapel) with a wooden, stone or a brass statue of the Lord Buddha with the bowls of water (Yonchap), light incense and butter lamps. The sacred mantra of Avaloketesvara, 'Om mani Padme hung', is recited several times apart from turning prayer wheels placed at the roadside and around the Chorten. At the community level, they organise regular religious prayers in the neighbourhood in which monks from the monastery are invited to recite holy texts. Also, the astrologer is consulted on the horoscope to ascertain the auspicious day, person's fortune, etc.

Even though polygamy said to be existed among Monpas, presently, monogamy is predominantly a followed practice in Mon land as practised by the Tibetan Buddhist. A clan within the tribal community plays the major role in regulating the marriage of the Monpa society. Earlier, the choice of alliances was confined to the identical Monpa group of Tawang but presently the younger generation have not only gone in accepting bride or bridegroom from other Monpa groups including from other tribes like Adi, etc. Incidentally, contrary to the freedom of choosing their desired partner, a popular culture in Northeast India, the Monpas have a limited freedom as still many Monpa families prefer the close cross-cousin marriage (both paternal and maternal).[3]

Special headgear, mask, yak skin shoe and traditional colourful woollen clothes are typically used among Monpa people during festival time, which is a stark contrast to the earlier generation of rural Monpas who used the attire for their day-to-day needs. The traditional dress of the male Monpa is *Kangnom* (short woollen) trouser or *Dhorna* (full-length woollen trouser) with *Toh-thung* (shirt) and *ali-phudhung* or *khanjar* (black woollen coat) called. Occasionally, a thick woollen red coat called *Chupa* or sleeveless coat called *Paktza* is also worn. On the other hand, the Monpa women wear *Shingka* (a sleeveless gown of light red colour with white stripes) and a red or black coloured square woollen cloth around the waist known as *Teng-ngakyima* with *baitoh-thung* (woollen coat). Besides, many Monpa women also wear the Tibetan dress known as the *honju* and a loose gown known as the *Chupa*.[4] The headgear worn by the Monpa differs based on the region of their stay. The '*Tsem-lham*' or traditional boots of the Monpa are made up of yak or cowhide soles and top with woollen cloth that reaches till the knees.

The locally grown vegetables and meats are often found stable food of the Monpas. Given Tawang's agro-climatic zone (temperate alpine zone), fewer crops like paddy, maize, millet and local pulse are produced at the few places which are one of the rural occupations of Monpas and also the third major sources of income of the people in the district. The Monpas mostly follow the traditional method of cultivation practices, including rain-fed shifting agriculture cultivation, which yields low productivity. The locally raised yak's milk is used for making home-made butter and dry cheese called 'churpi', salty local tea called butter tea is prepared with this cheese as one of the ingredients, and occasionally yak meat is used in their dishes.

But today with the development of modes of transportation, vegetables and meats such as beef, mutton, chicken and pork are imported from Assam, mainly from Guwahati.

## Food and Drinks Culture

The Monpas' food habits are mainly determined by the available animal husbandry and agricultural products, climatic conditions and religious practice. They are, by and large, non-vegetarian and eat beef, pork, yak meat, chicken, mutton and fish regularly. The meats are either collected by hunting from their surroundings or procured outside. The surplus portion of the meat is preserved, usually for winter, in dried form after cutting into small pieces. The meat is dried in the sunlight during summer and is stored in large aluminium or wooden boxes.

Furthermore, the vegetables, either dried or fresh, potato, cabbage, radish, pumpkin, eggplant, mushroom, spinach, chilli, etc., coupled with millet, rice, maize, barley, wheat and buckwheat are part of the Monpas' stable food.[5] They also occasionally consume a variety of fruits like apple, oranges, banana, peach, pears, persimmon, etc. based on the season.

The ethnic foods of Monpas are influenced by the Tibetan dishes, which dominate their food choices even today. A distinctive Monpa cuisine consists of rice or maize and millet paste with curry of boiled vegetables and cereals and meat. The main cuisines include common *Thukpa* (soup filled with noodles, minced meat and vegetables), *Thenthuk* (types of noodle soup), *Gyapa-Khazi* (vegetable rice), *Puta* (noodles), *Momos*, *Zan*, *Khura* (pancake) and *Bresi* (sweet rice). The Monpas use a lot of chillies in almost all their food as they are fond of spicy dishes given the cold weather. Also, the soup-based dishes apparently help the local Monpas not only to keep them warm, but also to hydrate.

Momos are another main dish that is consumed as a quick bite as well as the main course; the local variant is called Sha Momo (meat momo), Chura Momo (fresh cheese momo) or Patan Momo (vegetable momo). The milk, butter, ghee and *churpi* (cheese) from yak are other key ingredients that are found practically in all the Monpa foods including their major beverage butter tea. Butter tea, typical Tibetan style, is traditionally made by churning Mar (Monpa butter), salt and tea leaves in tall, narrow, copper or silver urns.

Besides, the locally prepared distilled liquor is generally called as Chang, Baang-Chang, Sin-Chang or Ara (also known as Arak) from maize, millet, barley, buckwheat, rice, etc. And it is an essential drink for every occasion of Monpas' lifestyle, be it birth, marriage, death, house warming, festival or any social occasions. These locally brewed liquors form an integral part of their food habit. On a regular basis, the local Monpas wash and boil the grains in a large vessel that were left to cool on a bamboo wad before mixed with powered yeast, then poured into a tightly lidded vessel and covered with things like blankets to ferment in underground cellars or home. Once fermented, they preserve it in aluminium vessels for days or months together. The most commonly available liquor is Ara or arrah (fermented rice wine).

Interestingly, today with the gradual influence of the western culture and the northern and eastern Indian effect with the tourist flow, Monpa community, particularly younger generation, has undergone significant radical changes in their diet habits. The commercially processed store-bought foods, fast foods like burgers and samosa, drinks like cappuccino coffees and commercial alcohols coupled with the intrusion of modern crop varieties have affected the traditional foods and drinking behaviour of Monpas. The materialistic lifestyle has, to an extent, intruded into the remote borderland community, thanks to the significant rise of tourist inflow, the presence of large number of security forces, availability of modern technology like internet, smart phones and the return of modern educated Monpas from other parts of India and abroad. As a result, there is a significant rise of diseases like diabetes, cholesterol, hypertension, cancer, heart disease and stroke, and other health disorders among the Monpas.

## *Eating and Drinking Habit*

So far as the eating habit of Monpa is concerned, generally at a time not all the members of the family take meals together. Each one has their own time, and their meal is based on his or her hunger and work. Usually, Monpas take their breakfast in a dawn before venturing out and dinner after the dusk. Lunch depends. Washing of hands with soap before meals is a rare practice among older ones and to an extent among the younger generation Monpa. Also, the plates are washed properly with clean running water. Sometimes, without cleaning the used plate is found used again by others in the family. But in the Tawang Town, the present-day food is being served in China clay plate or aluminium or stainless steel dishes.

The domesticated dogs are allowed to take their meal with human beings. The pets like dog are kept nearby while dining, and in some instances the pet too eats from their dishes on the floor or on their same plate. It is also observed that many of the Monpas do not cover the meal. It is also notable that fruits are directly consumed without further cleaning, which leads to health issues.

Similarly, Monpas drink water that is even not purified, either from the tap, lake/spring water directly or those stored in aluminium containers, bamboo tubes or gourd shells. Steel, glass or plastic tumbler glasses are used for this purpose. Noticeably, the drinking of water is least among the Monpa, as they consume local beer or butter tea time to time, in lieu of water.

But given the physical location of Monpas at high altitude site water that runs in the stream generally carries less pollutant, does not get stagnant and ultimately reduces the risk of getting polluted. Further, the other advantage is that air in highland is causally free from dusts and other pollutants.

## Health Issues

### *Lifestyle Diseases*

Monpa tribe in Tawang district used to pursue traditional food habits and depends on the Tibetan indigenous healthcare system. But due to gradual changes in lifestyle, they are witnessing diseases that are associated with modern lifestyle like hypertension nowadays. With the increasing sedentary lifestyle, the current generation of Monpas does not have much physical activity as compared to the elder generation and, hence, is highly vulnerable to health hazards and lifestyle diseases. The practice of taking yak for grazing, walking miles for commerce, etc., have reduced with the changing land pattern and increased security measures at the border. Besides, the harsh climatic condition forces them to stay indoors, which led to physical inactivity.

Most hilltop villages lack recreational facilities; even the Tawang Town has got proper stadium/ground only a few years ago. Also, most of the activities are scheduled only during the local festivals like Torgya. Otherwise, the current generation Monpa men are interested in watching television, playing cards and drinking indoor. The Monpa women are better placed at least as they are involved in household activities, managing the business/trade, rearing domestic animals, etc.

The effect of sedentary lifestyle is evident on younger Monpas with the prevalence of obesity, which is significantly higher in younger people with the higher risk of developing type 2 diabetes. According to the Indian Council of Medical Research–India Diabetes (ICMR–INDIAB) 2017 study, the Arunachal Pradesh has number of diabetes among the higher socio-economic status than other states. And the report highlights that the diabetes epidemic is in a state of transition and expected to further increase low socio-economic groups.[6] Ironically, many of them are unaware of family health condition as they do not have regular blood or health check-up and only depended on traditional medication, if and when required. Today, at least 70–80 per cent in Tawang town and about 30–40 per cent in rural areas know about the disease condition and some are aware of the symptoms of diabetes too.[7]

As per the Tawang district hospital administration, various kinds of lifestyle diseases are in raise, such as hypertension, cholesterol, gastritis and ulcer, respiratory diseases, tuberculosis, liver diseases and few cases of cancer like cervical cancer and lung cancer.[8] Apparently, the cases of hypertension are common owing to the high altitude and high sodium intake in food in the form of cheese, chilli, Chinese salt (MSG), salted preserved dry foods (salted and chilli, dried meats), etc. Today, there is an alarming statistic that 50 per cent of the Monpas are at risk of diabetes, compared to the national average of 23%. Incidentally, even the venerated monks and annis in the monastery and nunnery are also at higher risk of hypertension and diabetes. Further, the growing cases of gastritis and gall bladder stone problems owing to the intake of excessive spicy food combined with sodium are currently becoming quite common in this remote hill district. The Monpas are being urged to cut down on salt, chilli, cheese and alcohol by the medical professionals.

High liquor consumption, both home-made Ara (chung) and modern commercial liquors, has increased liver diseases like cirrhosis, colon cancer and renal failure in Tawang district. Additionally, the regular smoking and the use of drug made up of locally grown cannabis (in places like Lumla) and weeds by the Monpa youths are reportedly high, resulting in the rise of lung and liver diseases.[9]

## *Monpa Monks and Anni*

The Monpa monks and anni, who are studying and living in monastery and nunnery in Tawang and Bylakuppe, are too affected by the lifestyle diseases. With long hours of sitting in the prayer or meditation, chanting and Buddhist teaching without moving much around, the monks and annis in the monastery and nunnery are gradually facing health problems like diabetes, hypertension and cancer. Though they consume less food compared to a common Monpa, eating flatbread (roti) made of Maida flour (finely milled without any bran from wheat), high cheese an salted butter tea and consuming meat food and liquor have made them vulnerable to the diseases like high cholesterol, diabetes, hypertension, etc. Further, in the monkhood, they are generally not encouraged to venture outside for play, stretching and exercises. Particularly, about 50 per cent of (senior) monks are obese and hypertensive and also suffer from backache because of their sitting posture and a sedentary lifestyle.[10]

Other respiratory diseases such as asthma, chronic obstructive pulmonary disease (COPD), pulmonary fibrosis, pneumonia and lung cancer too are reported owning to tobacco smoking habits of Monpas.

The cases of gastrointestinal disorders among Monpas include conditions such as gastritis and haemorrhoids, and are reported commonly, as per the district health officials. The food and drinking habits are said to the major basis for it along with a sedentary lifestyle.

Apart from the lifestyle-related diseases, the Monpas face many communicable diseases, including tuberculosis. Importantly, tuberculosis has emerged as potential infectious disease at present in Tawang. As per the data from the District Medical Officer (DMO), 177 patients (4 indoor and 173 outdoor) were treated for tuberculosis and 81 (49 indoor and 32 outdoor) for Malaria at the government institutions in 2014–15 in the Tawang district.[11] About 1756 patients (881 indoor and 875 outdoor) were treated for fever during the same period.

As per the medical professionals at the District Hospital Tawang, sexually transmitted diseases like syphilis and HIV too are time to time reported. As per the DMO office data for 2014–15, at least 19 cases of syphilis were treated.[12]

## *Dermatological Disorder*

As Tawang district is situated at high-altitude region, which remains snowbound for nearly 5–7 months of the year, and the temperature remains low throughout the year, the prevalence of pediculosis and pityriasis simplex (dandruff) and occasionally cases of actinic dermopathy including hyperpigmentation are reported.[13] This also attributed to the lack of awareness of personal hygiene and cleanliness. In addition, as the Monpas do not take regular bathing owing to severe cold and the lack of modern geyser and electricity and have the habit of wearing the same unwashed clothes for months, contributing to dermatological problems.

## *Dental Problems*

No regular dental check-up and improper habit of brushing teeth have contributed to tooth decay and dental problems among many Monpas. Particularly, the cavity is found to be very common, and many cases of toothache are reported.

## *Women and Child Health Issues*

Significantly, Monpa women face various health issues due to early age marriage in rural areas that causes unfavourable physical effects on cohabitation and childbirth. The average age of marriage of Monpa women is 21 years in Tawang district. Particularly, anaemia and malnutrition are prevalent among pregnant Monpa women. Diabetes, gastritis and post-pregnancy-related illness are other health issues that Monpa women face. In terms of life expectancy at birth, Tawang has one of the lowest life expectancies of 49.79 years.[14] Evidently, people who live in upper hill ranges have lower life expectancies than who live in the plain. This is in comparison with the state level, which as per the 2005 human development report of Arunachal Pradesh is 55.05 years, one of the lowest in the country.

Similarly, the infant mortality rate (IMR) in Tawang is 3.45 per 1000 reported live birth (13 April to 14 March 2018), which is lower than state average of 33.[15] In fact, with the development of healthcare facility, IMR of Tawang has significantly decreased from 98 in 2005 to 7.04 in 2016 and 1 in 2018 (till April 2018).[16] The state average of IMR has also decreased from 77 to 33 per 1000 live births from 2005 to 2014. The main determinants of IMR are immunisation and access to the nearest healthcare facility among others. But till now, there is no special newborn care unit in Tawang district.[17]

According to the project manager of health department, who is in charge of district maternal and child health, family planning and adolescence health, the majority of the Monpa population in the district (50–60 per cent of 37,000) are dependent on district hospital for maternal health care. Also with the rampant Tantrism and superstition beliefs, pregnancies are not disclosed at early stage thereby causing delayed treatment and counselling by Integrated Counselling and Testing Centres (ICTC) of the health department.[18] Also, as per the district health department, free oral pills are available at hospitals and free condoms are distributed through vending machines for birth control. Yet, overuse of oral contraceptives is one of the main reasons of hypertension among young ladies/girls in the district.[19]

### *Mental Health and Addictions*

Almost 70 per cent of Monpa men are between the ages of 14 and 60 years, and less than 5 per cent among the Monpa women are used to tobacco chewing, whereas the prevalence of alcohol use is rampant among both men and women, particularly the local beer that is prepared in the household. And some 3–5 per cent of Monpa youth is used to locally grown wild weeds.[20]

### *Animal Bites and Stings*

Aside from the above health issues, as Monpas' settlement is often surrounded by hills and forests, animal bites from dog, snakes and scorpions are common. Notably, the cases of dog bite from the stray dogs have increased at an alarming rate in Tawang, particularly in Tawang Town. But reportedly, there has been no case of rabies-infected stray dogs in Tawang.[21]

### *Sanitation and Disposal of Sewage*

Another unhygienic factor that contributes to Monpa's illness is the issue of sanitation and sewage disposal system in Tawang. Noticeably, there are no proper drainage systems to canalise the sewage in the hill district. Many times, it mixes in the river or gets accumulated at the source level. This acts as a favourable breeding ground for mosquitoes and flies. Equally, there is no proper disposal of household or hotel refuse in a definite place. It is found to be heaped everywhere.

## Health-Seeking Behaviour

Health-seeking behaviour or healthcare-seeking behaviour (HSB) has been defined as 'any action or inaction undertaken by individuals who perceive themselves to have a health problem or to be ill for the purpose of finding an appropriate remedy'.[22] Notably, HSB was found to be low among the Monpa. However, the higher-income or educated Monpas in town in Tawang district have better HSB, particularly modern allopathic health care. On the contrary, the one who remotely settled in the hill or rural areas prefers to go with their family healthcare practice or religious healthcare practice and if severe illness, Tibetan form of medicine. Since consulting a qualified medical professional or finding private clinics, primary health centres and general hospitals during illness is very rare, even finding a chemist in the neighbourhood is difficult.

## *Traditional Healthcare Systems and Tibetan Medical Practices*

The Buddhism's influence on Monpa's practice of health care is evident in Tawang district. Buddhism stresses on personal responsibility and motivation as the precepts are training principles rather than commandments. Monpas are deeply religious and strongly spiritual. They hold the Buddhist teachings of great importance in their lives. In fact, the Monpas' family and the community observe auspicious Buddhist dates and restrict from having or selling non-vegetarian items on those days.

Tibetan medicine is inseparable from Buddhist principles of healthy living. It holds one's karma, past and present, responsible for any misbalance in one's body apart from causing disease, fear of evil spirits and germs. Further, it emphasises on the prayers and rituals which are believed significant for enhancing the efficacy of treatment and medicines. Consequently, the Monpas believe in the laws of karma; i.e. any action or deed that one does with their body, speech or mind will have a corresponding result. They believe in transmigration of the soul and reincarnation. And it is one's karma—positive or negative—that determines one's birth, illness or death. These strong beliefs are deeply entrenched in the way of life of the Monpas. Hence, Monpas often call the monks to recite sacred texts or consult them for medicines in order to cleanse the person from the negative karma.[23]

The Tibetan system of medicine that is preferred by many Monpas is in essence based on the Ayurvedic principles. It is based on the theory of the five elements, earth, water, fire, air and space. It is believed that these five elements interact with the key energies of the human body. The local traditional healer diagnoses the patients by examining the pulse and urine followed by asking the certain questions to patients for determining the illness. Subsequently, herbs, minerals and animal products are used in Tibetan medicine along with special diets prescribed. In line with the Tibetan medicinal treatment, vegetarian dishes are encouraged for consumption during illness.

In addition, if an individual's health allows, one day fast is recommended on new moon and full moon days. The Monpas generally have no problem taking any medicine prescribed by the traditional healer or Tibetan medical practitioner that helps.

In Tawang district, understandably as many as 90 per cent of rural Monpas prefer their own medication during illness while only 80 per cent of urban Monpas are reported to visit primary clinic or district hospital during their illness. However, 50–60 per cent pregnant Monpa women today are seeking medical care at the primary healthcare centre (PHC) than before.[24] The Monpas have sought to consult and seek medical assistance as the health facilities in Tawang are improved upon and made accessible to the people.

As an inhabitant of the eastern Himalayas that is rich in diversity of medicinal plant species, Monpas prefer their own system of health care to cope with diseases. The medicines are a conglomeration of methods, techniques, practices, cultural values, beliefs, traditions and ecological adaptations of them. In fact, the culture of traditional healing of diseases using medicinal plants (and animal ingredients too) that are grown in the region is still prevalent among them. Moreover, ritualistic performance is deeply attached in the Monpa society for curing diseases.

Given the Monpas' cultural, religious and the past political links with the Tibet, the Tibetan medicine is popular alongside their indigenous traditional curative measures. Fundamentally, the traditional wisdom of healing among Monpa community is passed down orally from one generation to the next by an elderly Monpa or traditional healers and spiritual gurus without much written document. Traditional healers are unqualified as per the modern education standard but possess the traditional know-how to treat the ailing using traditional methods. Besides, most of the traditional medical knowledge of the Monpas is culturally shared experiences. They believe that all the natural objects, including medicinal plants, are gifted by God for the well-being of the society. As a result, Monpas regard the forest and the environment as the abode of the spirits, and the people generally do not cut such trees or kill animals in general.[25]

The elderly Monpas have a rich knowledge of the wild grown medicinal plants in the region and its medicinal values, which are an integral part of their social life. They classify and utilise the various known medicinal plants, particularly herbs, barks, fruits and leaves of the trees, and plants, animals and other material ingredients that constitute the main source of essential ingredients as per the folk medicinal system. The Monpas refer to different parts of plants with different names; for example, tree is known as sheng, root as rabha, leaves as palap, flower as mento, seed as shu, fruit as shing-dho, bud as nyuyu, branch as allah, creeper as rui-khpo, trunk as sheng-dong, etc.

Some of the medicinal plants like *Aconitum heterophyllum, Asparagus racemosus, Ligularia amplexicaulis, Rhododendron hodgsonii, Swertia hookeri and Verbascum Thapsus* are used as an antidote against food poisoning, snake bite, insect bite and scorpion bite by the Monpas.[26] And edible plants like *Clerodendron colebrookianum* for treating high blood pressure, *Momordica charantia* for diabetes and mellitus, and *Lindera neesiana, Solanum etiopicum and Solanum indicum* for the treatment of intestinal parasitic worms like round and tapeworms are used.[27] Similarly, *Acorus colamus* (shuka) is used for blood dysentery, cough, gastritis; *Acceritu ferax* (zanto) for dysentery, poisoning and body pain; *Accerillum hetercphyvokles* (chanado) for body pain and cardiac arrest; *Acceritum heterophylium* (Bog-nga-karpo) for malaria, cough, snake bites, jaundice; *Azteseme tortuceum* (lomo sunchandawa) for tuberculosis; *Artemisia nilagirica* for wound healing and scabies; *Azadirachta indica* for stomach disorder and diarrhoea; *Clerodendrum colebrookianum* for hypertension; and *Rhododendron arboretum* for diarrhoea and throat infection.[28]

Another locally well-know medicinally important plant cum fungus *Ophiocordyceps sinensis* or *Cordyceps sinensis* (caterpillar fungus), a form of butterfly larva and fungus, is widely found in the Tibetan plateau, and one of the highly priced natural medicines in that region is used for kidney and heart problem and for treatment of infertility,[29] according to the Tibetan medicine.

Identification, deciding the edibility and processing of the plant are done by the Monpa community elders based on their wisdom. Sometimes, the Monpa women collect, grade and network with the market and sell these ethnobotanical products.[30] Since the ethnobotanical plants and local crops are not only the source of food, nutritional and health security for the local tribes, it also plays a major role in the household economy.[31]

But currently, this traditional medicinal knowledge is threatened by acculturation and deforestation. Importantly, with the adoption of cosmopolitan culture the younger generation Monpas lack interest in the knowledge of traditional medicine. The shift in the practices of the Monpas has also come with the exposure and feasibility of modern education, healthcare systems and lifestyles. With a more privatised economy, the community-led cultural and societal choices have altered. They prefer jobs that bring quick money and modern lifestyle. Furthermore, most of the Monpas lack the awareness about locally available medicinal plants. As a result, many medicinal value plant species that are available in the region, like *Ophiocordyceps sinensis* and *Paris polyphylla*, are endangered today. Ironically, some of them are smuggled to countries like China, Myanmar and Vietnam and reportedly a kilogram of the species can fetch up to Rs. 5000 to Rs. 200,000 in the international market.[32] For example, a medicinal plant of *Maxima Sp.* is sold at Tawang for Rs. 30/Kg and at commercial levels; it is sold at Rs. 8000/kg.[33] Today reportedly, the district forest office and NGOs like WWF are striving to document the medicinal plants in the district and sensitise the local tribes. In fact, a local Monpa named Tsering Topgey, traditional medical healer, has a medicinal plant nursery called Gomkhang Medicinal Garden at Gomkhang village, which is about 25 km from Tawang.[34]

## Government (Modern) Health Care

Presently, the Monpas of the Tawang district live in a pluralistic medical treatment option compared to a decade ago, such as both the traditional medicine and magico-religious performances prescribed by Tibetan doctors and allopathy medicine given by modem physicians or by the pharmacists obtainable. In many cases, multiple treatments are also practised by the Tawang Town Monpas. First, they seek a traditional home-based medicines, then Tibetan doctor and lastly for emergency cases physicians at the district hospital. The army hospital in the district is also treating the locals on the request of the Tawang district administration from time to time.

The demographic profile of the district elucidates this point of view. As the rural population comprises about 77.5 per cent of total, which is slightly higher than the state average 77.06%,[35] it is evident that the rural Monpa population in Tawang with inaccessible road and transportation prefer traditional methods to cure diseases first. The distance of healthcare centres from the villages and remote areas is apparently a major determining factor of overall state of healthcare system in Tawang. As per the census data, the maximum villages in Tawang district are not having primary medical facilities within 5 km or 5–10 km distance. Only in the bigger villages of above 500 populations, the availability of medical facilities is much better than that of village with lesser population.[36] But with the changing socio-economic dynamics, allopathic method has become more in practice among the educated and well-off Monpas.

According to the District Medical Officer (DMO), the number of Monpas seeking modern healthcare remedies has gone up in a decade, from about 25 per cent 15 years ago to 65 per cent today.[37] This is apparently due to the development of health infrastructure in the district. In a sense, the building and expansion of government district hospital and medical institutions in the district have immensely contributed in enticing modern healthcare practice among the Monpas. Twenty-five years ago, hardly any health infrastructure was available there, often necessitated long and difficult journeys to down Assam for medical treatment. Also, the provision of health care existed in a minor level based on military or civil outpost in the vicinity of the village.

As per the Tawang district portal data, Tawang district has totally 19 medical institutions, in which 22 are allopathic, one homoeopathic and one Ayurvedic institution. It is reported that there are 44 medical and paramedical personnel and 17 medical doctors in the district apart from 63 beds totally for the inpatients.[38] Also, Medical Mobile Unit (MMU) is set up at remote villages under the National Rural Health Mission to aid basic medical facility. However, the main source of treatment for the entire Tawang district, besides traditional method, is the District Hospital Tawang or *Khandro Drowa Sangmo*.

Out of 22 physical allopathic health infrastructures in Tawang, the health sub-centre is in majority in numbers among all the health service institutions in the district with two community health centres (CHC), six primary health centres (PHC), 14 primary health sub-centres (PHS), two maternity and child welfare centres, three family welfare centres, etc.[39] It indicates that some of the Monpas get usually health service mainly from the health sub-centres. At present, the district hospital has a footfall of about 200 patients in a day and most of them come for outpatient department (OPD). Besides the existing 50-bed government hospital, a new outpatient department is under construction with separate male and female wards and it is expected to be completed by 2018–19. Notwithstanding, Tawang districts have a very less number of beds, i.e. 1 per cent compared to rest of the districts in Arunachal Pradesh. Reportedly, the number of staff nurses available is about 30 and one nursing school with girls' hostel facility is available near the hospital premises.[40]

## *Human Resource Position*

The District Hospital Tawang has one ophthalmologist, two gynaecologists, one health department project manager and six general duty medical officers to deal with an inflow of around 180–200 daily outpatient department (OPD) patients. Also, the basic pathological facilities, including ECG, ultrasonography (USG), X-ray and basic blood tests, are available at the District Hospital Tawang. In addition, one allopathic pharmacy with one head pharmacist and two associate pharmacists is there in the hospital pharmacy. And the whole district itself is reportedly having only around ten medical shops. Evidently, medicines are available in subsidised rate under the Prime Minister's Jan Aushadhi Yojana. Yet, drugs related to oncology and psychological disorder are not available in the Tawang pharmacy. According to some patients, the expired medicines are sold in some chemist in the district.[41]

But with no surgeons and certain specialist including anaesthesiologists (currently, one of the doctors is trained to administer anaesthesia by the WHO with the collaboration of Government of India) coupled with the lack of MRI scan at the district hospital, in case of any major health problem, the DMO approaches the nearest hospital at Tezpur, Itanagar or Guwahati or the army field hospital. Due to the bad shape of the road from Tawang to Guwahati, which is around 500 km long, several cases of critical patients dying on the way too are reported. In this regard, the construction of a road via Bhutan to Guwahati as demanded by many Monpas will reduce the travel time as it will reduce the distance by approximately 100 km.

## Conclusion

The Monpas of Tawang in Arunachal Pradesh have the sociocultural and religious setting that is quite similar to that of Tibet and Tibetans. The socio-economic changes are gradually taking place in this remote and once isolated Monpa land with Monpa youths going outside for higher education and much interested in cosmopolitan culture, business and politics that can offer swift power and money. The changes are not only noted in the economic conditions but in the health sector too. Earlier only traditional health care was predominant, but today the modern allopathic remedies are sought by many Monpas with the development of district health hospital and government health institutions. In fact, many are now more dependent on government district hospital and army hospital in case of emergency nowadays. But still, in the rural upland Monpa settlement people largely depend on home remedies and traditional healthcare system such as Tibetan medicine. This can be related to the lack of accessible road and transportation to visit the faraway health centres or district hospitals. Some with better socio-economic conditions with own transportation visit the district hospital for treatment.

Regardless of whether one opts for traditional or modern healthcare remedies, one of the interesting aspects is that the Monpas believe in karma as per the religious custom that physical happenings have links with deities and spirits. The Monpas by and large suffer from common flu (cold, cough, fever), skin-related diseases, including itches and rashes, gastritis and ulcer problems and intestinal worms, jaundice, diabetes, hypertension, tuberculosis and liver disorder owing to climatic condition and intake of salt, yak cheese, chilli and the country liquor in large quantity.

In addition, it is evident that the Monpas are not conscious of personal hygiene which too is partly responsible for the prevalent diseases. Irregular bathing, tooth brushing and washing clothes and utensils are perceptible in contributing to many illnesses. Further, the existing living conditions of Monpas in Tawang district with no proper drainage and sewage disposal system are too responsible for some of the diseases and ailments. The lifestyle changes and health problems among the borderland population have serious implications for the country's socio-economic development and health security.

The elder Monpas in a family or community and traditional healers including the Tibetan practitioner use different materials ranging from their natural and non-natural surroundings to treat various illnesses. Notably, in the Monpa society, one cannot isolate the illness and healing from the environment as their livelihood is closely linked with environment. Consequently, ethnobotany or ethno-medicines play a significant role in treating many diseases. The Monpa of Tawang is one among the tribes in Arunachal Pradesh having vast knowledge of medicinal plants that are grown in their surroundings and their uses in different diseases. Sometimes, the complicated diseases are treated with ethno-medicines coupled with prayers and rituals as suggested by the local priest.

The traditional healing system of Monpas that has played a very important role in the healing of different diseases of their tribes in remote hilly areas is passed from one generation to another through oral narrative; some of them are preventive in nature, whereas others are curative. They have learned it through traditional experience of trial and error method. Currently, there is a choice of healing and treatment options for Monpas. For curing diseases, the Monpas opt for traditional medicine or resort to Tibetan medicine as their first preference, and only for some cases like delivery complication they prefer modern health care.

Undoubtedly at the moment, many traditional health practices of Monpas are facing the danger of extinction. The global trend of commercialisation of ethno-medicinal biodiversity and insensible nature of younger generation Monpas has endangered the locally grown medicinal plants. There is an urgent need to pay adequate attention in the conservation and documentation of medicinal plants grown in Tawang district and ethno-medicine practices. The health scenario among Monpas warrants urgent and effective measures in arresting the raising of health issues and enhancement and distribution of medical facilities including both allopathic and alternative healthcare measures in the district.

## Endnotes

1. Refer, Tamalika Chakraborty, Somidh Saha, and Narendra S. Bisht, "First Report on the Ethnopharmacological Uses of Medicinal Plants by Monpa Tribe from the Zemithang Region of Arunachal Pradesh, Eastern Himalayas, India," *Plants* (Basel). 2017 March; 6(1): 13.
2. Ved Prakash, *Encyclopaedia of North-East India*, Volume 3, Atlantic Publishers & Dsitributor, 2007, p. 1208.
3. Veeranki Maheswara Rao, *Tribal Women of Arunachal Pradesh: Socio-economic Status*, New Delhi: Mittal Publications, 2003, p. 179.
4. Refer Tawang Tourism, *Dress & Ornaments*, https://tawangtourism.in/culture/dress/.
5. http://tawangtourism.in/culture/cuisine/.
6. Refer Ranjit Mohan Anjana et al., "Prevalence of diabetes and prediabetes in 15 states of India: results from the ICMR–INDIAB population-based cross-sectional study," *Lancet Diabetes Endocrinol* 2017, at https://www.thelancet.com/journals/landia/article/PIIS2213-8587(17)30174-2/fulltext.
7. Analysis from the author's field visit to Tawang district in April 2018.
8. Author's interaction with the DMO, District Hospital Tawang, April 8, 2018.
9. Author's interaction with the health workers and NGOs, Tawang, April 10, 2018.
10. Author's interaction with the Medical Officer at Sera Mey Monastery health centre, Bylakuppe, Karnataka, 18 July 2018.
11. Author's interaction with the duty doctors in District Hospital Tawang, 13 April 2018.
12. Author's interaction with the duty doctors in District Hospital Tawang, 13 April 2018.
13. Author's interaction with the duty doctors in District Hospital Tawang, 13 April 2018.
14. *SHDR Survey and Sample Registration System Bulletin, April 2002* quoted in Human Development Report 2005, Government of Arunachal Pradesh, at https://www.undp.org/content/dam/india/docs/human_develop_report_arunachal_pradesh_2005_full_report.pdf.
15. Refer *State profile of Arunachal Pradesh (2013–2014)*, Regional Resource Centre for North Eastern State, Ministry of Health and Family Welfare, Government of India, at http://www.rrcnes.gov.in/state_profile/Arunachal%20Pradesh.pdf.
16. Refer *Human Development Report 2005*, Government of Arunachal Pradesh, at https://www.undp.org/content/dam/india/docs/human_develop_report_arunachal_pradesh_2005_full_report.pdf.
17. *Arunachal Pradesh - State Profile*, at http://www.rrcnes.gov.in/state_profile/Arunachal%20Pradesh.pdf.

18. Author's interaction with Dr. Canlee Dorjee, the Health Department Project Manager at the District Hospital Tawang, 13 April2018.
19. ibid.
20. Author's interaction with the DMO, District Hospital Tawang, 8 April 2018.
21. "Tawang strategy to tackle stray dog menace," *Arunachal Observer*, June 21, 2019 at https://arunachalobserver.org/2019/06/21/tawang-strategy-tackle-stray-dog-menace/.
22. Editorial, Health Seeking Behaviour in Context, *East African Medical Journal*, February 2003 p. 61 at https://www.ajol.info/index.php/eamj/article/viewFile/8689/1927.
23. Refer "Department of Tourism Tawang, A Complete Guidelines," at https://cdn.s3waas.gov.in/s39b70e8fe62e40c570a322f1b0b659098/uploads/2018/07/2018070315.pdf.
24. Author's interaction with the duty doctors in District Hospital Tawang, 13 April 2018.
25. Khongsai, M, Saikia, S P & Kayang, "Ethnomedical Plants Used by Different Tribes of Arunachal Pradesh", *Indian Journal of Traditional Knowledge*, Vol. 10(3), July 2011, pp. 541–546.
26. Tsering J., Tag H., Gogoi B.J., Vijay Veer, "Traditional Anti-poison Plants Used by the Monpa Tribe of Arunachal Pradesh," in Vijay Veer, Gopalakrishnan R. (eds), *Herbal Insecticides, Repellents and Biomedicines: Effectiveness and Commercialization*, Springer, New Delhi, 2016.
27. Namsa ND, Mandal M, Tangjang S, Mandal SC., "Ethnobotany of the Monpa ethnic group at Arunachal Pradesh, India," *Journal of Ethnobiology and Ethnomedicine*, Vol. 7 (31), 2011 at https://ethnobiomed.biomedcentral.com/articles/10.1186/1746-4269-7-31.
28. Refer Nima D Namsa et al., "Ethnobotany of the Monpa ethnic group at Arunachal Pradesh, India," *Journal Ethnobiol Ethnomed*, 2011 at https://ethnobiomed.biomedcentral.com/articles/10.1186/1746-4269-7-31.
29. "Caterpillar Fungus: The Viagra of The Himalayas," *NPR*, October 9, 2011, at https://www.npr.org/2011/10/09/141164173/caterpillar-fungus-the-viagra-of-the-himalayas.
30. Umeshkumar L. Tiwari1, Amit Kotia and Gopal Singh Rawat, "Medico-ethnobotany of the Monpas in Tawang and West Kameng districts of Arunachal Pradesh, India," *Pleione* 3(1): 1–8, 2009 at https://www.researchgate.net/publication/263315753_Medico-ethnobotany_of_the_Monpas_in_Tawang_and_West_Kameng_districts_of_Arunachal_Pradesh_India.
31. Refer Craig Jeffrey, "The 'Viagra' transforming local economies in India," *BBC News*, July 7, 2012, at https://www.bbc.com/news/magazine-18735544.
32. The information was gathered from the field visit and interaction with the District Forest Officer (DFO), Tawang, and some local Monpas, April 2018.
33. Dr. Piyush Dutta, Fellow WWF in his interaction with author at Guwahati, 22 April 2018.
34. Author's interaction with Dr Abdul Qayum, the District Forest Officer (DFO), Tawang, 9 April 2018.

35. Refer Census of India 2011, District Census Handbook Tawang, Series – 13, Part XII-A.
36. Refer Census of India 2011, District Census Handbook Tawang, Series – 13, Part XII-A.
37. Author's interaction with the DMO, District Hospital Tawang, 8 April 2018.
38. https://tawang.nic.in/health/.
39. District Census Handbook, Tawang, 2011.
40. Author's interaction with the DMO, District Hospital in Tawang, 8 April 2018.
41. Author's interaction with patients at the District Hospital Tawang premises, 8–10 April 2018.

**M. Mayilvaganan** is Associate Professor in the International Strategic and Security Studies Programme at National Institute of Advanced Studies (NIAS), Bengaluru. He was a former visiting senior lecturer at the Department of International and Strategic Studies, University of Malaya, Malaysia, where he taught South Asia in International Relations, Modern Warfare and International Politics and formerly Associate Fellow at the Institute of Defence Studies and Analysis (IDSA), New Delhi. Mayil holds a PhD from the School of International Studies, Jawaharlal Nehru University (JNU) and has spent time as a researcher/independent scholar in Canada and the US. His research interests include strategic and security issues concerning India and its neighborhood, Indo-Pacific, foreign policies of India, China, Japan and the US, Borderland and non-traditional security issues.

# Chapter 7
# Natural Resources and Biodiversity Conservation Practices in Tawang

**Nasima Khatoon**

## Introduction

Arunachal Pradesh, the largest state in the northeast of India, is spread over 83,743 square kilometres with nearly 79.96 per cent of forest cover (ISFR 2017). The geographical area of Arunachal Pradesh constitutes 2.55 per cent of the total area of the country, but it constitutes 10 per cent of the total forest cover in the country. Two of the world's 35 biodiversity hotspots—the Himalayas and the Indo-Burma—fall within the state with endangered species of flora and fauna. It is the largest mountainous state of India and recognised as a globally important ecosystem. It comprises over 5000 species of plants which are distributed along one of the largest elevation gradients in the world, ranging from lowland tropical forests to alpine vegetation in the Himalayan Mountains. Most of the forest cover in Arunachal Pradesh is designated as unclassified state forest. The forest is grouped under six major categories, i.e. tropical, subtropical, temperate, sub-alpine and alpine vegetation, secondary forest and aquatic vegetation, each comprising subtypes primarily based on altitude and climate change.

The Tawang district of Arunachal Pradesh is spread over 2172 km$^2$, and 54.19 per cent (State Forest Report 2017) of the total area is covered by very dense to open forest. The population density of the district is just 23 per square kilometres, and the forest area is largely under the control of the local community. The local Monpa community has been dependent on the forest for centuries for their daily needs and has also been able to conserve the ecologically fragile forests. Hence, pressures on the local biodiversity were negligible until a decade ago. Presently, several factors have affected this ecologically fragile yet naturally conserved area. These include rapid development of tourism industry and infrastructure development, human interference

N. Khatoon (✉)
National Institute of Advanced Studies (NIAS), Bengaluru, India
e-mail: khatoon.nasima15@gmail.com



M. Mayilvaganan et al. (eds.), *Tawang, Monpas and Tibetan Buddhism in Transition*,
https://doi.org/10.1007/978-981-15-4346-3_7

in the form of change in lifestyle, extensive extraction of natural resources mainly wood to support the rapidly growing hospitality industry and various other purposes.

In addition, these activities not only disturbed the largely sustainable way of life of the locals but also the rich biodiversity of the eastern Himalayan region. This is quite evident from the State Forest Report (2017) which has recorded 190 $km^2$ decline in the state forest cover, which is more than twice of the 74 $km^2$ decline in state forest cover recorded in 2011. While at the Tawang district level, forest cover has reduced significantly by 49 $km^2$ or two per cent in 2017 as compared to 2011, which is more than the reduction in the total state forest cover of 0.54 per cent in the same time period.

The mundane existence of the Monpas is dependent on the local biodiversity from which they draw their daily subsistence from food, clothing to shelter. In turn, the Monpas have carefully knitted a practice of conservation and sustainable living into their religious worships and culture in order to keep their dependency on the nature harmonious. But presently, the increase in natural resource extraction owing to the change in lifestyle has bordered on destructive impact on the biodiversity.

Hence, it has become imperative to have an assessment of the current status of the biodiversity, existing conservation plans and the local community's dependence on forest resources so that an evaluation of the various levels of threats and working plans for the management and conservation of these forest resources is conceivable.

## A Biodiversity Profile of Tawang

Tawang lies in the westernmost part of Arunachal Pradesh and is bordered by Tibet in the north, Bhutan in the southwest and West Kameng district in the east. The border district with incredibly large diversity of plants, insects and reptiles has an average elevation of 2669 metres (8757 ft). Habitat types (Mishra et al. 2004) found in the district are rhododendron forest above treeline area of 12,000 ft, high-altitude alpine and sub-alpine meadows above 4000 m altitude, temperate oak and conifer forest from 2000 to 4000 m and subtropical broadleaved forest ranges up to 1800 m.

As per the state forest report, Arunachal has suffered a huge loss of forest cover and in India approximately 2254 $km^2$ medium density forest (MDF) has turned into non-forest land in past two years. Arunachal's forests account for 1/3rd of habitat area within Himalayan biodiversity hot-spot. The state is habitat for over 5000 species of plants, about 85 species of terrestrial mammals and over 500 species of birds, apart from a large number of reptiles and insect species.

In the Tawang district of the 2172 $km^2$ area, 1177 $km^2$ is covered by forest, which is 54 per cent of the total geographical area. According to the state forest report of 2017, this includes 341 $km^2$ of very dense forest, 448 $km^2$ of moderately dense forest and 388 $km^2$ of open forest (Fig. 7.1). This forest is an important repository of terrestrial biodiversity of Tawang and plays a key role in influencing the socio-economic and cultural attributes of the Monpa society.

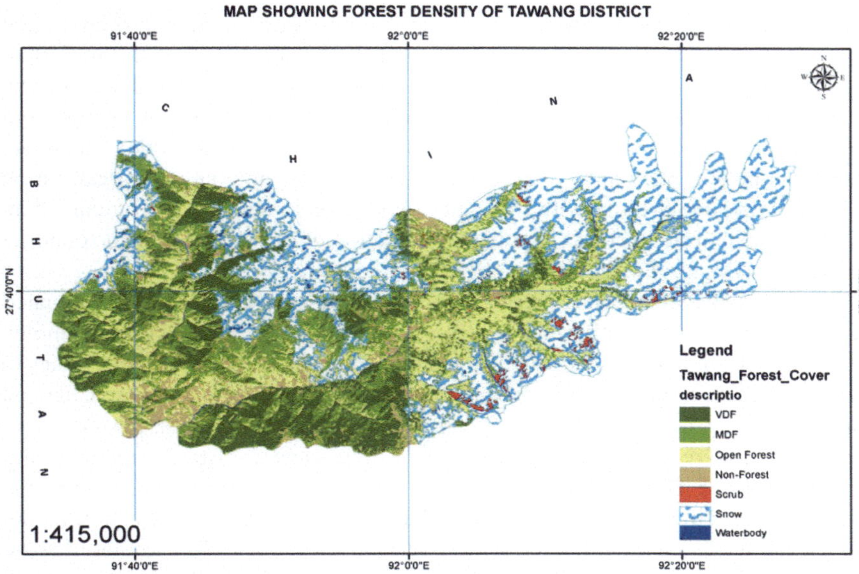


**Fig. 7.1** Map of forest density of Tawang district. *Source* Tawang Forest Division, Department of Environment and Forest, Government of Arunachal Pradesh, 2018

The report estimated loss in 25 $km^2$ of very dense forest and 38 $km^2$ of moderately dense forest and slight increase in open forest. Shifting cultivation and developmental activities are said to have caused the loss, although logging is not mentioned. While Tawang's total forest cover, i.e. 54 per cent of total geographical area, is significantly lesser than the state forest cover, i.e. 79.96 per cent and lowest in terms of other districts of Arunachal Pradesh, this loss in forest cover is an alarming trend.

The forests in Tawang are commonly considered to be one of the best preserved and richest in biodiversity. A major portion of these forests are "sacred forests" and community forest land. The vegetation and the wild life change with the changing altitude of the Tawang forest. The subtropical zone ranges up to 1800 m, and it is dominated by mixed broadleaf forests. It is also home to diverse wildlife species like Himalayan goral, Chinese pangolin, common leopard, etc.

Temperate zone ranges from 2000 to 4000 m of altitude; it is dominated by mixed conifer, oak, rhododendron and diverse shrubs and herbs. Tawang is also home to the rarest of rare animals like red panda, serow, Mishmi takin, Chinese goral, etc. High-altitude wetlands are found above 3000 m altitude in temperate, sub-alpine and alpine zones. Most of these wetlands are concentrated in lake complexes like Nagula Wetland Complex, Bhagajang Wetland Complex and Pangchen Wetland Complex of Tawang. Besides sheltering rare flora and fauna and harbouring fragile ecosystem, these high-altitude wetland complexes support major river systems including Tawang Chu and Nyamjang Chu.

The Nagula Wetland Complex[1] in Tawang comprises more than 100 lakes, and the complex acts as the reservoir for the Nyamjang Chu. Some of the important lakes of this wetland complex are Kheset Tso, Kyalem Tso, Khamkhar Tso, etc. The catchment area of these lakes is a rich repository of flora such as Primula and Rhododendron, and also provides habitat for fauna such as the ruddy shelduck, Himalayan monal, pheasants, finches and accentors. The international organisation World Wildlife Fund-India with the state forest department and the Indian Army has played a key role in conservation and protection of these lakes and surrounding biodiversity (Deputy Forest Officer 2018).

The Bhagajang Wetland Complex comprises 12 major lakes and is situated between the altitudes of 4000 and 4400 m in the district. This wetland complex is the reservoir of Tawang Chu which is the main river in the district and tributary to the Manas River. Some of the important lakes of the wetland complex are Changna Dorjee La Tso, Chandrezig La Tso, Gyalpo Namse, Dorjee Phamu, etc. All these lakes are of high religious significance to the Monpas and the Buddhist community all over the world. The catchment area is rich in different varieties of flora and fauna, flora species such as Primula and Rhododendron and faunal species such as the Chinese goral, snow leopard, serow and red panda. The area is also habitat of some important avifauna like the ruddy shelduck, blood pheasant, Himalayan monal, snow pigeon, etc. The state forest department and stake holders from the civil society such as the Tawang Monastery are working jointly with WWF-India for the conservation and protection of these lakes and surrounding flora and fauna.

The third wetland complex of the Tawang district is the Pangchen Lumpo Muchat Community Conservation Area (CCA) wetland complex that lies between the altitudes of 4000 and 4500 m in the northwestern part of Tawang district. The wetland complex comprises seven main lakes, and these lakes act as a reservoir of the Nyamjang Chu River. This river supports the winter habitat of the vulnerable black-necked crane in India, which is a religiously sacred bird for the Monpas. Some of the important lakes of the complex are Burma Tso, Gosung Tso and Oma Tso. The floral species at the catchment area are Rhododendron, Aconitum, Bistorta, Primula and other grass species; some of these plants are rare medicinal plants such as Aconitum, Rheum nobile and Saussurea gossypiphora. The wetland complex is also a habitat of fauna such as musk deer, snow pigeon, Himalayan Monal and some amphibian species. The state forest department and residents of two nearby villages, Lumpo and Muchat, work jointly with WWF-India for the conservation and protection of these areas.

Sub-alpine and alpine zones range from 4000 m and beyond; the sub-alpine habitat is dominated by shrubs of rhododendron, juniper and diverse medical and aromatic herbs. The alpine zone only supports sparse growth of vegetation. One of the main characteristics of this zone is that it supports highly adaptable and rare species like snow leopard, blue sheep, musk deer, blood pheasant, etc. One of the noticeable features of pristine forest of Tawang is forests of Rhododendron sp. can be found even around 14,000–15,000 ft, which is above treeline area of 12,000 ft.

## The Factors Affecting the Biodiversity in Tawang: A Critical Understanding

### *Collection of Timber and Firewood*

In December 1996, the Supreme Court (SC) of India made a landmark ruling after hearing a Civil Writ Petition (TN Godavarman vs. the Union of India and others) alleging state failure to control logging, and the SC verdict on the case included an interim order prohibiting logging without government permission. Initially, the case concerned Tamil Nadu and later because of the reported large-scale loss of forest, the northeast of India was also included. The SC order of 1996 states that "all ongoing activities in any forest...without prior approval of the central government, must cease forthwith".

Many experts (Karlsson 2005) have argued that this Supreme Court order disregards the sixth schedule that is the right of autonomous district councils on full jurisdiction of all forests, except the reserved and protected area. In Tawang, as about 60 per cent of forest land falls outside the protected area; hence, according to the judgement the forest produce of this unprotected forest could also be regulated by government. This in turn has severally affected the relative freedom and rights of the indigenous tribe on the natural resource management of the forest land.

In case of northeastern region of India where shifting (*jhum*) cultivation is prevalent, the use of the term "forest" has been inadequate. The very nature of jhum cultivation is that the trees and shrubs reclaim the land that has been cultivated. Hence, the difference between the farmer's use of land and large-scale timber extraction by contractors needs to be understood within the control and management of land and resources in "all forest" areas. The centralised state control over the forest lands and resources constrains the relative freedom of indigenous tribal people who depends on the forest for livelihood. This questions the underlying assumption that the state institutions and not indigenous tribal people are best suited to conserve and manage forests and resources in a sustainable way and hence contribute positively in conserving the ecosystem.

Besides this, there have been cases of illegal timber trade by outside promoters with large corruption and money flows down to Assam or Tezpur, as these cities are the two main hubs of business, road connectivity and works as lifeline for Tawang district. According to official data, about 82 $m^3$ of illegal timber was seized during 2005 to 2010 and no forest encroachments are reported so far. The shortage of timber for over-exploitation of forests has led to escalating prices and attractive profit, which makes this trade even more lucrative. This in turn has led to increase in illegal felling of trees, and political intervention for granting licences to the plywood and sawmill has gave rise to great deal of corruption.

The Arunachal Pradesh government has adopted a policy of generating employment by dealing out contracts for road construction. Of the state's total plan allocation (Roychowdhury 2015), the Public Works Department (PWD) gets 20.14 per cent. Another 38.4 per cent goes for transportation and communication. The Monpas in

Tawang depend largely on this contract work for livelihood as it generates easy income. Hence, due to the availability of fast returns from contractual work over daily wage and labour, the need to live out of nature in a self-sustainable way is slowly fading away mainly from the Monpa life.

While it is important to develop infrastructures in this difficult terrain, the system, based mostly on contracts and government sanctions, is getting increasingly corrupt. Selection of sites for roads and measurement of angles for earth cutting on fragile slopes are often not in accordance with legal stipulations.[2] As a result, roads get constructed in ecologically unsuitable areas.

On the contrary while the ban on the extraction of timber is for the protection of the forest, question of cutting of firewoods for daily consumption has not been addressed. On the way to Tawang and in the district, large amount of firewoods can be observed piled up in almost every house. On the occasion of unavailability of firewood trees, Monpas depend upon mainly oak and rhododendron trees for their everyday needs. Oak is used for this purpose as it is believed to generate considerable heat. But oaks are important for water retention and for keeping the mountains cool. Hence, excessive use of oak for firewood will gradually affect the moisture retention capacity of the mountainous soil, leading to erosion and subsequently contributing to the climate change of the region. The effect of climate change is visible in the region. Bomdila, the main city of adjacent West Kameng district, frequently misses snowfall. Similarly, rhododendron wood is banned for using as firewood, but villagers sometimes use it unaware of its ecological significance.

High-altitude rhododendrons of Tawang contribute largely in the biodiversity growth of the region; indiscriminate use of these trees for firewood has led to slow extinction of many unique species (Paul et al. 2005). As at higher elevation soil density is relatively lesser and plant growth rate also decreases with higher altitude, these high-altitude plant species take relatively more time to grow. The high demand of firewood with less possibility of rhododendron replenishment has also led to habitat loss of some of the very unique fauna species of this region. In mandala of Tawang due to large-scale deforestation, the species of red pandas have become vulnerable and a simultaneous shrinking of forest areas with more human interference has further led to the decrease in panda population with frequent human–animal conflict.

With the increase in urbanisation and budding tourism industry, the demand for firewood has manifested. Hence, lack of alternate plans has led to irreparable damage to the fragile ecosystem. Therefore, a more comprehensive and inclusive environmental strategy should be considered to maintain the ecological balance in one of the most enriched biodiversity hotspots in India.

## *Loss of Grazing Lands*

The Brokpa, the nomadic pastoral community of the Monpa tribe, is known to rear sheep and yak for their livelihood. Brokpas, pastoral community of Monpas, visit alpine meadows (4000–4500 m) for six months before migrating down during winter months. The loss of forest cover is posing a serious threat to the fodder and energy resources of the Monpa villages. In and around the Monpa villages, in Tawang, the local Monpa herders can be seen travelling long distances to graze their undernourished cattle. One of the alarming trends in the present time is that the nomads leave for the grazing lands early, before the onset of summer in Tawang, and stay in the higher elevations for longer period before coming back to lower elevation during the wintertime. This indicates the slow change in weather conditions and relative increase in overall temperature and hence longer summer period. This change in overall temperature and grazing pattern affect the yaks adversely, which has led to significant reduction in its population. The villagers have noticed gradual increase in skin diseases and high mortality rate of yaks as a result of this changing summer months.

Similarly, due to changes in the climatic conditions and increase in temperature, the high-altitude grazing lands are also becoming limited. Being a strategic borderland, a large tract of Tawang's grazing land is already occupied by the stationed security forces for construction of roads, bridges and bunkers. This has caused immense hardships to the local communities. Earlier, the herders could move freely across the Himalayan borders with their animals which are not allowed anymore owing to heavy militarisation of the border area.

In recent years, however, the number of households involved in pastoral activities has declined in view of the changed perception and modern education which is leading to adopt new livelihood opportunities and living style. The difficulties in accessing higher grassland for grazing with accidental straying into the quarters of land owned by the Indian Army could also have contributed to this change.

## *Hunting of Wild Animal and Avifauna*

Monpas are followers of Gelugpa sect of Tibetan Buddhism and a strong believer in life after death. Monpas do not hunt animals due to this religious significance, but the religion allows them to use dead animal. Hence, they have incorporated a unique way to hunt animals—they chase the animal to the edge of mountain so that the animal falls to death and they can use it thereafter. In the recent decade, this practice has become more prevalent. Simultaneous destruction of habitat[3] by Monpa due to collection of large amount of firewood, infrastructure development, increase in population and adoption of a unique blend of modern and traditional lifestyle have led to not only changes in lifestyle and dilution of traditional cultural values but in turn witnessed a rise in illegal hunting and wildlife trade. Sometimes, wild animals plunder agriculture fields for food, and with human settlements growing close to the forested land the incidents of human–animal conflict have manifold.

Similarly, *Brokpas* are now often forced to climb the mid-elevation and high elevation areas for livestock rearing and grazing. But this has now become another reason of livestock predation by wild animals. Although conservation of biodiversity and sustainable utility is deep rooted in Monpa culture, they hunt various avifaunas.

While these act as one of the major sources of protein for them, eventually more hunting due to high demand has depleted the population of these birds, triggering an imbalance in the biodiversity of the region. Reportedly, over 60 types[4] of birds are hunted by Monpa people.

## *Forest Fire*

Forest fires are a serious threat to the hill slopes of this Tawang region. The occurrence and intensity of forest fires are high during October to April, when the long dry season coupled with wind spreads the fires to larger areas. Maximum percentage of fire incidences occur in the lower elevation (<1500 m), and from that point of view Tawang is relatively less affected by forest fire. However, forest fire spreads faster in uphill; thus, movement of fire is relatively high in steeper slope (Ahmad et al. 2018). They can be caused either naturally through lightening and rubbing of dry bamboos and stones or by people accidentally or intentionally.

In Tawang, people practise, clearing and burning forests as a part of shifting cultivation system (Gadgil and Meher-Homji 1985), where land is cleared for growing crops by first clearing of trees and vegetation and then burning it thereafter which is an old practice that still persists in relatively smaller scale in Tawang than other neighbouring districts of Arunachal Pradesh. Shifting cultivation has been identified as major cause of forest degradation both qualitatively and quantitatively in Arunachal Pradesh. Local communities having poor socio-economic conditions who live along the fringes of the forest continue to burn forests to support their livelihood by producing fodder and agriculture (Ahmad et al. 2018) and to facilitate the collection of non-timber forest products. Topography, weather and availability of inflammable material are various factors which decide the intensity of forest fires.

## *Demand for Autonomy and Control Over Ecosystem*

Compared to other parts of India, the tribal communities in the Northeast India enjoying provisions under the Indian constitution's Sixth Schedule are often seen as being in a favourable position. Yet, even if it is said to be correct, an interesting fact is that more or less all tribal or indigenous people from this region speak of their lack of control over land and resources and thus argue for increased autonomy and "self-determination". Autonomy movements have thus dominated the political scene in Northeast India (Karlsson 2005).

Tawang is no exception. Development and modernisation have finely altered the close-knitted tribal structure of power which has in turn led to an identity assertion among the Monpas, mostly led by the monks.

The debate for autonomy gained traction in Tawang once the infrastructure development and large-scale utilisation of natural resources other than the community slowly started affecting the ecological and sustainable way of life of the Monpas. The autonomous debate led by the monks as the cultural and social representation of the Monpas has often emphasised that the recognition of Tawang as an autonomous region will give the community greater freedom in determining their land ownership and use of bio-resources, thereby controlling the economy of the district. The authority over forested land can impact sustainable management of natural resources based on the use of Monpa's traditional knowledge of biodiversity conservation.

Autonomy would also help them decide the extent of the present infrastructure developments; most contested of them are the two proposed mega-hydropower projects in Tawang in 2006 and 2007, respectively: one on Nyamjang Chu and other on Tawang Chu rivers (two phases). The Monpas have often considered the Ladakh autonomous region as pathway to preserving their own control over their nature. According to the local Monpas,[5] the *Rinpoche* (principal monk) has noted that due to its autonomy in Ladakh, the roads and infrastructure developments are better in the region. The debate supporting autonomous council also suggests that the rampant corruption during the transfer of fund from central to the state level and then to the district will be relatively much lesser as the fund from central will be directly be sanctioned to the district.

## Conservation Process

Forests are important repositories of terrestrial biodiversity and have always played a key role in influencing socio-ecological and cultural attributes of human societies including livelihood activities of traditional societies living in the area.[6] In the contemporary time, there has been increased realisation that biodiversity conservation is interlinked with the traditional communities inhabiting in the forested landscape. Involvement of these people is crucial in the conservation process.[7]

The people of Arunachal Pradesh depend on forests for their livelihood since time immemorial. More than 60 per cent of these pristine forests in Arunachal Pradesh are under community custodianship. In these forests, Monpas and other tribes exercise their traditional rights of collecting fuel-wood, timber, forest products and medicinal plants, which form the source of their livelihood. Since in Tawang, a substantial section of forest cover is under the ownership of local people, effective and sustainable management of the rich biodiversity invariably necessitates community participation which will also contribute towards economic betterment of the Monpas. However, population growth, gradual changes in Monpa lifestyle and rapid unplanned infrastructure development have in many places led to the over-exploitation and destruction of forest areas.

In this context, various measures to conserve the biodiversity have been undertaken by local community, state government, WWF-India and some non-government organisations. Although these initiatives have helped to minimise the biodiversity destruction in the region, more comprehensive models can be undertaken.

The Monpa community of Tawang manages land and water resources through their traditional culture and belief system. The village council is called Mangma, and it is headed by Dzongpon (governor of the tribal village council). The Dzongpon operates within its jurisdiction around the village, the Dzong (province), and deals with all kinds of major disputes related to forest and water resources such as dispute over forest land holding, jhum-linked land holdings, non-timber forest product harvests, dispute over grazing land and issues related to district government and the villages. All other issues are taken care of by the village headman called *tsorgen* or *gaonburah* through a collective decision-making process organised at the Gumpa or monastery.[8] Although monastery's opinion is highly valued in Monpa society, monastery does not have influence on internal decision-making of village council.

For the management of forests, Monpas have distinct categories of forests—clan forest (Tsanpasa, Tsan), privately owned forest (Rangasa), community forest (Mangsa/Borang), oak-dominated forests having a special place (Ja-sesing). Well-defined rules ensure sustainable use of the forest resources, for example, if a person who is from the same village but from a different clan needs to extract forest resources strictly for local consumption and prior permission is easy to obtain. Outsiders may be allowed to use forest resources on the basis of negotiated monetary transactions. Impact of religion is seen in every aspect of sociocultural life of the Monpas. Sustainable biodiversity conservation is closely linked with their unique rituals and custom.

In the Zemithang Valley, a single mountain is considered to be of religious value for the deity Shockseng Gompa, and therefore, extraction of plants from this hill region is avoided. Similarly, Zemithang has many religiously important sites around it, such as villages like Kalaktang and Kharman, with the twelfth-century built Gorsam Chorten Buddhist Stupa near the bank of Nyamjang Chu River. Because of the religious significance of the area, villagers strongly protest government proposal for construction of mega-hydropower project. Adverse effect on local ecosystem including change in the course of water flow of Nyamjang Chu River and large-scale habitat loss of some of the rare fauna species of the region are main reasons behind strong anti-dam sentiment of the local people. The Zemithang Valley is the habitat of red panda, Himalayan musk deer and Arunachal macaque, among other species. All of these species has been listed as "endangered" by International Union for Conservation of Nature (IUCN) Red List of threatened species (The IUCN Red List of Threatened Species n.d.), because of over-exploitation these species are facing a very high risk of extinction in the wild.

The place is also the breeding ground of the migratory black-necked crane, which has immense religious and ecological value for the Monpas. The habitat loss due to infrastructure development has been one of the main reasons which led to the protest movement in 2016. Hence, when it comes to ecology–community relation among Monpas, the religious dimension plays as an important factor in helping to conserve certain species of ecologically important flora and fauna in the region.

Similarly, because of the religious sentiment linked to the species of *Rhododendron, Cryptomeria, Juniperus* and *Cupressus,* these flora species are protected by the

Monpa community. *Juniperus* sp. is also considered to be sacred and used for incense-making. The religious value linked to this species could be determined from use of the timber by the Monpas for the construction of the "Chosum",[9] where the Lord Buddha is placed. This is also evident from the biodiversity-linked rich traditions embedded within the Monpas that what is socially valued invariably has an ecological and economic value too and these values linked with biodiversity are indeed enabling factors towards sustainable conservation and management of biodiversity.

Although, presently a trend can be observed where erosion of traditional conservation practices is slowly leading to the depletion of the natural resources of the region. Religious argument of conservation is being refuted by the younger generation. Unlike the relatively older generation of Monpas who sustained their livelihood based on forest and forest-based products, the younger generation is mostly involved in occupations that are based on outside sources. Especially, the Tawang Town and adjacent villages have easy inflow of cash owing to compensation of land by army, blooming tourism industry and also due to it being the strategic border town. Due to availability of these relatively easier means of earning, the need to live out of nature and self-sustenance is being gradually given up by the Monpa society. Hence, it can be observed that a self-sustainable society is gradually becoming dependent on outside sources for their livelihood.

In an attempt to solve this issue, WWF-India has been working in the region since 2004 with the member of local Monpa community for the development of a tourism industry that helps in conservation of the rich biodiversity and environment. In this respect, WWF has initiated Community Conservation Area (CCA) project, which is for the community to voluntarily safeguard the natural resources and use them sustainably.

In Tawang, three Community Conservation Areas (CCAs) have been established for this purpose by the WWF-India, the Pangchen Lumpo Muchat CCA in Pangchen Valley, the Pangchen Lakhar CCA and the Pangchen Socktsen Diksum CCA. These CCAs cover approximately 215 $km^2$ in Tawang district for the conservation purpose.

The major ongoing programmes in the CCAs aim at promoting ecotourism, protecting the black-necked crane and its habitat mainly in Zemithang Valley, safeguarding the future of the red panda and its habitat, and introducing BioLite HomeStove in order to reduce the amount of firewood used in daily activities. Similarly, to reduce the use of ecologically important and vulnerable plants, WWF-India encourages local communities to grow trees solely for firewood. A CCA Management Committee comprising local villagers responsible for the ecological conservation and socio-economic development of the village has been set up for conducting regular wildlife surveys and patrolling the demarcated areas to ensure that no illegal activity takes place.

The CCAs also help community members in developing long-term and sustainable livelihood opportunities like community-based ecotourism. In an ecologically fragile and pristinely beautiful place like Tawang, tourism should flourish but community-based tourism must be encouraged to preserve the ecology and unique culture of the tribal community. For this purpose, state government with the collaboration of WWF has initiated community-based ecotourism in relatively remote places like

Zemithang Valley of Tawang. The Directorate of Tourism has initiated various training programmes focusing mainly to train unemployed youth and villagers to become tour guide and housekeeping stuff, and to start business of transportation, homestays and hotels.

Zemithang has six such homestays, where government officials regularly conduct training programme. Lumpo, Kalaktang and Kharman also have four such homestays. Under Chief Minister's *Paryatan Vikas Yojana*, the homestay employees are educated and given loans to install proper toilets, cookery training like baking, food production and front office training. This has proved to be a highly inclusive initiative providing high economic return and also helped in tourism sector as tourist inflow has increased significantly. For example, while the total number of tourist visited Tawang in 2016 was 24, 702, in 2017 it was 43,294.[10]

Apart from this, WWF-India with the help of state government and local community has contributed significantly in conserving high-altitude wetlands of Tawang. These wetlands have immense ecological value and are crucial to sustain the biodiversity of the Himalayan region. These wetlands also support river systems of the area and have transboundary significance. Presently, WWF-India and state forest department are developing a management plan for four wetland complexes in the district, i.e. Bhagajang high-altitude wetland, Nagula high-altitude wetland and Pangchen Lumpo Muchat CCA high-altitude wetland. State forest department with the help of Monpas, Indian Army and WWF-India is conserving these areas and surrounding flora and fauna species. Conservation activities like clean-up drives, awareness programme and proper disposal of garbage (which is increasingly becoming a problem in the absence of sewage treatment plant) are also being conducted. Apart from this, the state forest department has set up Nature Interpretation Centre and Rhododendron Park in the Tawang Town to increase awareness regarding the local biodiversity and for in situ conservation of some vulnerable and ecologically important flora species. High-altitude rhododendrons of Tawang are another area where large-scale conservation work is needed to protect many rare species from being extinct.

## Conclusion

The Monpas of Tawang have traditionally developed the culture of conserving and managing nature and natural resources. These traditional conservation practices relating to spiritual and valued species by the Monpa community are to sustain the continually evolving cultural practices, while conserving the rich biological diversity that they value. When traditional values have contributed towards conservation of biodiversity, the evolving changes in eco-cultural values from within and outside with the passage of time have posed threat to community-based sustainable development.

While change is unavoidable with new technological developments, the challenge lies in enabling this traditional community to have a traditional conservation practice linked with sustainable developmental pathway, based on a value system that they appreciate and therefore participate and work towards making the region ecologically viable.

In the recent past, there have been attempts to build two mega-hydropower dams, which have been rejected by the Monpas. It is obvious from this discussion that the Monpas indeed do have a rich set of cultural traditions linked with conservation of biodiversity. However, the efforts of the governmental agencies to develop infrastructure in the name of "modernising" oftentimes tend to lead to social disruptions, with serious adverse consequences, which should be avoided at all costs.

For sustainable environment-friendly development of the region, the local communities should be taken into confidence as large part of the forest land in this region is under community control and community participation is vital in protecting the biodiversity hot-spot of the area. Therefore, conserving the rich cultural diversity and traditions are the key to conserving the biological diversity. The two being intertwined with one another it can contribute significantly towards community-linked developmental pathway, which will help the community to embrace modernity while still practicing environment friendly rich tradition and culture.

## Endnotes

1. Information collected during the author's visit to Nature Interpretation Centre, Tawang, 6 April 2018.
2. Interaction with local people and contractors at Tawang and Bomdila.
3. Interview with Dr Piyush Dutta, former landscape coordinator for Western Arunachal Pradesh, WWF-India, Guwahati, 22 April 2018.
4. Kripaljyoti Mazumdar and Prasanna K. Samal, "Conservation, Management and Hunting of Faunal Resources among Monpas and Sherdukpens in Arunachal Pradesh, Eastern Himalaya," In Ramakrishnan, P.S., Saxena, K.G., Rao, K.S. and Sharma, G (eds.), *Cultural Landscape The basis for linking biodiversity conservation with the sustainable development*, UNESCO, NIE, New Delhi, 2012.
5. Views documented during a field visit to Tawang, 11 April 2018.
6. Herrmann, T. "Indigenous knowledge and management of Araucaria araucana forest in the Chilean Andes: Implications for native forest conservation." Biodiversity and Conservation 15: 2006, 647–662.
7. Alcorn, J.B. "Indigenous people and conservation." *Conservation Biology* 7:424–426, 1993.
8. From the interview with Nawang Chota (Gaonburah of Lumpo village in Zemithang, Tawang) during field trip in Tawang, April 8, 2018.

9. Mihin Dollo, Prasanna K. Samal and Darwin Megejee. "Biodiversity Linked Value Systems of the Monpas and Sherdukpens of Arunachal Pradesh." In CULTURAL LANDSCAPES: The Basis for Linking Biodiversity Conservation with the Sustainable Development, by K.G. Saxena, K.S. Rao, G. Sharma (Ed.) P.S. Ramakrishnan, 59–81. New Delhi: United Nations Educational, Scientific and Cultural Organization, 2012.
10. Tourism information officer, Tourism Department, Government of Arunachal Pradesh, Tawang, 2017.

## References

Ahmad, F., L. Goparaju, and A. Qayum. 2018. Himalayan forest fire characterization in relation to topography, socio-economy and meteorology parameters in Arunachal Pradesh, India. *Spatial Information Research* 26 (3): 305–315.

Deputy Forest Officer. 2018. Tawang, in an Interaction with the Author. April 11, 2018.

Gadgil, M., and V.M. Meher-Homji. 1985. Ecology and management of World's Savannas. In *Land Use and Productive Potential of Indian savanna*, eds. V.M. Tothill & J.C. Mott, 107–113. Canberra: Australian Academy of Science.

Karlsson, Bengt G. 2005. Indigenous Natures: Forest and Community Dynamics in Meghalaya, North-East India. In *Ecological Nationalisms*, eds. Gunnel Cederlof, and K. Sivaramakrishnan, 170–198. New Delhi: Permanent Black.

Mishra, C., A. Datta, and M.D. Madhusudan. 2004. *The High Altitude Wildlife of Western Arunachal Pradesh: A Survey Report.* CERC Technical Report No. 8. Mysore, India: Nature Conservation Foundation, International Snow Leopard Trust and Wildlife Conservation Society (India Programme).

Paul, A., M.L. Khan, A. Arunachalam, and K. Arunachalam. 2005. Biodiversity and conservation of rhododendrons in Arunachal Pradesh in the Indo-Burma biodiversity hotspot. *Current Science* 89 (4): 623–634.

Roychowdhury, Anumita. 2015. *Chopping Down the Future,* 07 June 2015. Accessed January 12, 2019. https://www.downtoearth.org.in/coverage/chopping-down-the-future-29723.

State Forest Report. 2017. *Forest Survey of India, Ministry of Environment and Forest.* http://fsi.nic.in/isfr2017/arunachal-pradesh-isfr-2017.pdf.

The IUCN Red List of Threatened Species. n.d. Accessed June 24, 2019. https://www.iucnredlist.org/.

**Nasima Khatoon** is a Research Associate in the International Strategic and Security Studies Programme at National Institute of Advanced Studies (NIAS), Bengaluru. Her areas of research include nuclear diplomacy, national security, and West Asia.

# Chapter 8
# Infrastructuring Arunachal Pradesh Borderlands: A Case of Tawang Borderland

**Mirza Zulfiqur Rahman**

## Introduction

The province of Arunachal Pradesh in India's northeast has often been described as the land of the rising sun.[1] The conflict with China in 1962 forced India to take deeper notice of the significance of the northeast as a critical frontier in its national security calculations. Chinese troops had advanced as far as Tezpur, and India was on the defensive regarding critical infrastructure required for faster troop deployment. After the war was over, India pushed towards building a basic-level artery system of roads and military infrastructure on its borders.[2]

At the same time, however, New Delhi was wary of developing a strong infrastructural presence in Arunachal Pradesh and, till the end of the past decade, had followed a deliberate policy to neglect the development of Arunachal Pradesh and parts of the upper banks of the Brahmaputra in Assam, lest Chinese troops roll down the hills again.[3] The roads built immediately after the 1962 war were only targeted as *cosmetic development* geared towards meeting India's troop deployment needs and inadequate to fulfil an all-round development and cross-border trade for the people in Northeast India. Even for the targeted troop deployment purpose, India clearly lagged behind China[4] as the latter made rapid strides in building infrastructure all along its critical border points, especially in Tibet,[5] in sync with its Western Development Strategy through the 1990s.

Discussing the diverse types of border contestations in a region, van Schendel (2013) depicts the India–China border as a 'sensitive border' marked by uncertain sovereignty and apprehensive territoriality. He goes on to argue that such a border cannot be called a border at all, and in official parlance it is referred to as the 'Line of Actual Control' (LAC) or more famously as the 'McMahon Line'. It is based on the ground presence of the respective militaries along the border or what is the

M. Z. Rahman (✉)
Institute of Chinese Studies, New Delhi, India
e-mail: mirzalibra10@gmail.com

 
M. Mayilvaganan et al. (eds.), *Tawang, Monpas and Tibetan Buddhism in Transition*,
https://doi.org/10.1007/978-981-15-4346-3_8

perception of the borderline by them, usually negotiated on a regular basis by 'long-range patrols' which perform 'area domination exercises' and monitor and inspect border pillars, some perhaps once in a year or two, given the remoteness of the borderline. The way ahead, according to van Schendel, is to approach such borders from the lens of 'anthropology of frayed edges' than with the definite 'geography of lines'.[6]

Borrowing from the formulation of van Schendel (2013), it is exactly at these 'geography of lines' at the Sino-Indian border in Arunachal Pradesh, which the nation state looks at, both from the Indian and the Chinese sides, and also lies the 'anthropology of frayed edges' which is the worldview of communities inhabits inhabiting these borderlands of Arunachal Pradesh in India and in the Tibetan Autonomous Region in China. The hydroelectric dams that the state government of Arunachal Pradesh has planned, with as many as 168 MoUs being signed by them with various public and private dam-building companies, threaten the worldview of local communities along the borderlands. This is manifested in the anti-dam protest movements and mobilization among local communities inhabiting Arunachal Pradesh, with a view to protect their livelihood means and sacred spaces, from the western parts led by the Monpas, to the central parts led by the Adis and the Galos to the eastern parts led by the Idu Mishmis.

## Situating Arunachal Pradesh Within the Indian Nation State

The province of Arunachal Pradesh, which was named in 1972 as a Union Territory within India and gained full statehood in 1987, has an approximate territory of around 83,743 km$^2$. This territory was earlier termed as the excluded areas of the province of Assam under the Government of India Act 1935 and subsequently as the North-East Frontier Tracts (NEFT) and then as the North-East Frontier Agency (NEFA) till the year 1972. This history of this frontier region in the Eastern Himalayas is marked by varying degrees of continuities and discontinuities, in terms of both Indian and Chinese claims over the territory given the long-standing Sino-Indian boundary dispute, India's engagement with its frontier space and the development discourse formulated by New Delhi for Arunachal Pradesh over time.

The colonial construct of the territory of Arunachal Pradesh remains the point of continuity in which New Delhi looks at the space post-independence. This is the result of the discontinuity that we see between the developments in the frontier state in comparison with the rest of Northeast India and furthermore the rest of India. As Baruah (2008) points out, the central tenets of the colonial wisdom about the region laid the foundation of the colonial spatial order, that is, the administrative boundaries and the rules of inclusion and exclusion that prevailed. The very fact that the 'inner line', which was a colonial construct to determine the inclusion and exclusion of economic and development space in Northeast India, runs to this day, speaks volumes on the continuity of the colonial spatial order in India. The tea and oil industry in the plains of Assam was the core economic zone for the British, where all development took place, with no interference with the 'inner line'.

Robert Reid, while emphasizing the area's heterogeneity, remarked that the people in these borderlands share one thing in common and that is they did not have any affinity with the people of the plains, or the people of India proper, be it historically, linguistically, culturally or racially, and that if they were tacked on as an Indian province, it was only a matter of historical accident and a natural administrative convenience.[7] Further from this point, we can observe that the various tribal identities in Arunachal Pradesh had very little historical, cultural, racial or linguistic links, and the space of the province was marked by New Delhi as a matter of administrative convenience, carrying on colonial legacy, and this is somewhat true even today in the face of poor intra-Arunachal connectivity.

As Dzuvichu (2013) argues that the politics of access in borderland spaces in Northeast India that plays out in terms of drawing territorial boundaries, road-building exercises, recruitment of labour, control over the flow of commodities and the introduction of new technology have led to the transformation of polity and society and the reconfiguration of identity. This aspect of the colonial times can be witnessed in post-colonial India, where the politics of access to space in various parts of Northeast India, including Arunachal Pradesh, especially through the act of 'road-building', has happened over several 'waves' of developmental attention of New Delhi, each with a particular strategy of control over the frontier areas and its borderland people in mind. In spite of these discontinuities in the development dynamics in post-colonial Northeast India, Jacob (2015) describes the region as one of 'coexisting civilizations', instead of 'clash of civilizations', and as a region where tribal communities have historically resisted the presence of the state and its affiliation and control,[8] the trends of which are variably visible in Northeast India today.

The Bengal Eastern Frontier Regulation of 1873, a colonial construct, which marks Arunachal Pradesh, Nagaland and Mizoram with its inner line rules and regulations, where outsider's and non-tribals have restrictions to settle or buy land and property, has been the 'way of life' for common Indian's engagement with this frontier space. In the two-way interactions between Arunachal Pradesh and the rest of India, the obtaining of an inner line permit has been the norm and considered 'business as usual' which is a continuity of the meaning of the given space that the province of Arunachal Pradesh holds. The Indian imagination of Arunachal Pradesh as a frontier land, a remote borderland, is further reinforced on a regular basis, by these processes of inclusion and exclusion.

This has been severely tested in recent times, when there has been an increase in connectivity and infrastructural projects in Arunachal Pradesh, necessitating a sizeable flow of daily wageworkers and manual labourers to the state, including an increased size of the military forces. Arunachal Pradesh is fast witnessing a change, which is unprecedented in size and scale in terms of what has happened in Northeast India in the past, through these road-building and infrastructure projects. The tribal communities of Arunachal Pradesh, who have lived in relative isolation for many decades, have to now grapple with such changes, and the politics of access has changed the social, cultural and more importantly the intra-Arunachal tribal dynamics leading to changes in local politics. The sort of upheavals caused by these rapid changes is seemingly working under the same norms and institutions that have governed these spaces for many decades, when such activity was less. Today, it has caused political and social tensions, including protests, blockades and other repercussions of community mobilization, including violence, which has been witnessed in region before.

As Sanjoy Hazarika in his book 'Strangers of the Mist' describes the 'Thousand Years in a Lifetime' analogy in the context of political and social changes in Nagaland, the same holds true for Arunachal Pradesh, albeit in much shorter a time frame, in its interaction with the Indian state and the rapid movement of development and capital flow. Some have described Northeast India as remote frontier by virtue of its unique history and geography, with the political management of this sensitive region being a delicate and difficult task,[9] while others have described this region as a 'resource frontier', with the politics of nature defining a space, and looking essentially from the perspective of the periphery rather than the perspective of the state-centric institutions and mode of thinking.[10] Both of these aspects are in play in the changes that we witness in Arunachal Pradesh today, where traditional livelihoods and imaginations are ruptured by the influx of development capital from New Delhi.

There are attempts in Northeast India of social, cultural and political mobilizations of communities, which seek to construct a self-identity of a space, which is distinct from the one that is constructed by the state, be it the colonial or the post-colonial state, and which becomes a concern for the latter in its project of achieving legitimacy and development in that given space.[11] The framing of the border space and the local political discourse by the various tribal communities in Arunachal Pradesh, vis-à-vis the disputed nature of the given territory between China and India, plays into the dilemma of New Delhi which in turn leads to the imposition of legitimacy through big-ticket infrastructure projects and border road-building plans. The space of Arunachal Pradesh, from an Indian strategic perspective, has moved from a scorched-earth policy to a huge inflow of development and infrastructural projects, in about a decade, eliciting local social and political responses/mobilization.

## Scorched Earth to Development Corridors: Local Perceptions of Infrastructure

Various tribes inhabiting Arunachal Pradesh have, over the past decades, been introduced to the processes of integration to the Indian nation, and it can be termed as largely successful.[12] The emotional component associated with national integration in a large nation such as India has been very strong in some parts and very weak in other parts of Arunachal Pradesh. The challenge for New Delhi in its attempts to integrate Arunachal Pradesh into the national mainstream is two-pronged, first the pulls and pressures of development priorities within the seven states of Northeast India, and then the ability to integrate Northeast India into the national mainstream.[13]

The varying degree of integration within Arunachal Pradesh can be attributed to the aspects of language, education, local perceptions of development in the state and the development of roads and connectivity. The western, central and eastern parts of Arunachal Pradesh are connected in disparate ways to the rest of India, thereby creating disparity also in terms of national integration among the tribes of the state. While none of these parts can claim to be better connected to India, they are also not mutually connected within Arunachal Pradesh due to the lack of proper intra-Arunachal connectivity, which the Trans-Arunachal Highway Project, when completed, will be able to remedy. Incidentally, two decades ago, Arunachal Pradesh was described as a region, which had not awakened to modern civilization. There has been relatively more development in the other northeastern states, Assam, Meghalaya, Mizoram, Manipur, Nagaland and Tripura, except Arunachal Pradesh. The people of Arunachal Pradesh did not get the level of developmental attention from the Indian government, which they deserved all along, because of the scorched-earth policy.

New Delhi now wants to assert its legitimacy in Arunachal Pradesh by the initiation of developmental projects and thereby integrate it better into the national mainstream or at least into the Northeast India mainstream. This is termed as coming a bit too late, but it is nevertheless encouraging for the future of the state and its people. However, New Delhi has demonstrated over time that it seeks to conduct its policy in Arunachal Pradesh with a considerable amount of diplomatic caution and that it does not take even symbolic postures on the most controversial issue in Sino-Indian relations.

Developmental measures announced by New Delhi hold some promise, but it has to be seen if the big power projects really benefit the local people. The many hydroelectric power projects in Arunachal Pradesh include some which would be India's highest capacity projects in terms of actual output projected and its overall potential, for instance, the Subansiri, Siang and Dibang rivers in the state. The increased frequency of helicopter services for civilian use is also in the offing, and several disused airfields in the state are being revived, for both civilian and military use. Only two states in Northeast India, Arunachal Pradesh and Meghalaya, do not have a proper functioning airport. The only available option is the skeletal helicopter services provided by the state government, which have seen accidents in recent years,[14] leading to serious safety concerns among the people of Arunachal Pradesh, which has seen several accidents in the past. The upgradation of the Lilabari Airport in Lakhimpur, Assam, which was active in the 1980s, can cater to the people of central Arunachal Pradesh and connect the capital city of Itanagar better with the rest of India, which must be put on the fast track by the government. Itanagar was linked with the railway network in Assam in April 2014, with the 20-km-long Harmuti Naharlagun railway track completed, the first passenger train service started.

## Sacred Spaces and Infrastructure Development in Arunachal Pradesh Borders

In order to illustrate the community scale of understanding borders and connectivity, the local tribal interactions with the Sino-Indian international border are employed. The Mishmi tribe consists of the Idu Mishmis who are most numerous, followed by the Digaru Mishmis and the Miju Mishmis who collectively inhabit the Mishmi Hills of eastern Arunachal Pradesh, consisting of the districts of Lohit, Anjaw, Lower Dibang Valley and Upper Dibang Valley.[15] There are Mishmi villages across the LAC in Medog and Zayu (Chayu) county of the Tibet Autonomous Region (TAR) of China, who belong mostly to the Miju Mishmi subgroup and are known as the Deng. There is no interaction between the Mishmi villages in India and China except for the Mishmi shepherds and hunters who frequent the hills and forests close to the border. The Meyor ethnic community, who are Buddhists, inhabit some villages along the LAC in Anjaw district, live by farming and hunting, and have family and relatives across the border in China.

Hunting is common among the Mishmi as it is connected to their animist religion and traditions, which require wild meat to appease the spirits and protecting deities in their festivals and family offerings. The Mishmi hunters often spend weeks in the forests along the Sino-Indian border and come in contact with Chinese hunters who often cross into the Indian side. Apart from consumption and religious needs, hunting in the Mishmi hills is also done for commercial purposes, where musk deer and black bears are hunted for their pods and gall bladders, respectively; these are sold to businessmen from mainland India or across the border in China for the traditional medicine market.[16] The invocation of the hard border in the Sino-Indian border at the national level in India comes through as quite soft in the easternmost sector of this border, where such interactions regularly happen, not necessarily through defiance by the community, but by the nation state looking away.

The number of Mishmi traditional priests, locally known as 'shamans', who conduct the animist customs in community festivals and family ritual offerings, has fallen significantly, and there are almost none in the Mishmi villages on the Chinese side. Consequently, the Chinese side woos Mishmi priests from the Indian side with good honorarium offer to conduct such rituals. These points towards a unique international border exchange rituals, offerings and animist belief systems between the Mishmi tribal people living in China and India. Mishmi villagers in Arunachal Pradesh say that they know that their Mishmi brethren cannot openly follow their animist religious rituals and practices on the Chinese side due to the prevailing communist ideology. Therefore, they accord a high value to cross-border interactions between shepherds, hunters and traditional priests, in order to preserve their common Mishmi animist religion, culture and identity, depicting common socio-spatial community belonging.[17]

The construction of hydropower dam projects in Arunachal Pradesh has been seen as violating scared spaces of local communities in the borderlands, be it Buddhist sacred spaces or traditional animist sacred spaces, including that of the Monpas, the Adis, the Galos and the Mishmis. Particularly in western Arunachal Pradesh, in the Tawang sector, the construction of the Tawang Chu and the Nyamjang Chu hydropower dam projects has increased apprehensions among local Buddhist Monpa population, about the environmental impact and the larger inundation of Buddhist sacred sites. This is where the 'anthropological frayed edges' come into focus in an examination of the developmental politics of this strategically located frontier region of India. The framing of New Delhi of these hydropower dam projects is of the ushering of development as well as strategic positioning of its sovereignty rights over Tawang. The local Monpas are more concerned about protecting the sacred sites they revere.

Monks and locals in Tawang, birthplace of the sixth Dalai Lama, have held regular protests against proposed dams in the district in March 2013, which they feel will threaten over 450-year-old Galden Namgyel Lhatse monastery and local people's livelihood.[18] The environmental impact of these dams in the Tawang sector on the black-necked cranes, which are locally revered as the reincarnation of the Sixth Dalai Lama, who was born in Tawang, makes for the rallying point behind the protests by Buddhist monks and local Monpas over the past few years, which saw three people killed in police firing in 2016. The Nyamjang Chu hydropower project is on the Nyamjang Chu River, which is a 780 MW project and the dams on the Tawang Chu River are in two phases of 750 MW each, known as Tawang I and Tawang II. These dam projects will severely impact the habitat of the black-necked crane, an endangered species, whose habitat extends into neighbouring Bhutan.

Arunachal Pradesh can be easily divided into three distinct geographical parts, the western, the central and the eastern parts of the biggest state in terms of territory in entire Northeast India. These three geographical zones come from a function of the lack of interconnectedness, meaning the routes of infrastructural connectivity do not smoothly join these three parts within the state of Arunachal Pradesh. The rivers that criss-cross the Eastern Himalayas make for the challenges of internal connectivity in Arunachal Pradesh, which the Trans-Arunachal Highway Project seeks to address. This highway project starts in the westernmost parts of Arunachal Pradesh, from near the tri-junction of China, Bhutan and India in the Tawang district, and ends in the easternmost fringes of the state, near the tri-junction of China, Myanmar and India, where the rolling hills of Dehing Patkai ranges begin.

The infrastructure development is prioritized in the westernmost districts of Arunachal Pradesh, which comprises of the East Kameng, the West Kameng and the Tawang districts, and distinguished by the geographical features of the meandering Kameng river that flows through the Eastern Himalayan landscape. The western part has been significant in the political, economic and cultural life of the state of Arunachal Pradesh, because of a variety of reasons, primarily because of the religious significance of Tawang, Bomdila, Dirang and Rupa. This apart, the incidents of the 1962 Sino-Indian war including Tawang being the most prized possession in the control of India, which is disputed by China and is now part of the Sino-Indian boundary talks and the importance of Tawang as the most popular circuit of tourism in Arunachal Pradesh are reasons of importance to Arunachal Pradesh.

The road-building infrastructural development in the western part of Arunachal Pradesh holds immense strategic priority because of the Sino-Indian relationship. Additionally with the initial phases of the Trans-Arunachal Highway Project being concentrated here, and given many Chief Ministers of Arunachal Pradesh have hailed from the western region, particularly from the Monpa community, many of these projects have been fast-tracked in recent time. Of the Chief Minister's hailing from the region is also the charismatic leader Dorjee Khandu, who was killed in a tragic helicopter crash in 2011 in Tawang, and the current Chief Minister Pema Khandu. Many strategic infrastructural projects have been announced in the western part of Arunachal Pradesh, including the tunnel through the Se La Pass, which was announced by New Delhi in 2018, and initial surveys are already underway.

The road construction activities and for that matter any interventions in the Kameng, Tawang Chu, Nyamjang Chu rivers and its many tributaries, in terms of hydropower interventions, are prone to many challenges, which are unique to the Eastern Himalayan landscape and besotted with the development lag that the region, and all of Northeast India for that matter, has witnessed in the past five–six decades after India's independence. This was when the conscious scorched-earth policy was being followed in Arunachal Pradesh, and this has slowed down many of the projects, which have now been announced by New Delhi, in the past decade of infrastructural development impetus seen in the region. The Tawang borderland has a very tough construction landscape, and it does throw up new geological surprises[19] often, similar to the Himalayan landscapes further to west and north.

The borderlands between Tawang and Bhutan have been the sites of traditional pilgrimage routes, with a considerable number of monks from Bhutan living in monasteries in Tawang and studying religion. The sacred spaces between Tawang and Eastern Bhutan cannot be confined into the nation state boxes of India and Bhutan, respectively, as we understand it in the strategic sense of political control and mobility. The sacred spaces that are going to be affected by the various hydropower infrastructure projects being planned by India in Tawang are going to affect downstream Bhutan as well, as the rivers flow from India to Bhutan. This is a classic upstream–downstream riparian conundrum and contestation that the region is witnessing due to planning and announcement of many dams in the fragile Himalayan landscape, considered sacred to the local communities straddling these international borders.

The appropriation of sacred spaces for large-scale infrastructural developments has seen popular protests in Tawang in the past, there were two among the three people killed in police firing incident in 2016, and a tenuous peace is now prevailing in the region. The framing of infrastructural interventions in borderlands such as Arunachal Pradesh, and even more so in the strategically important region of Tawang, is primarily in the national strategic security perspective, and therefore public participation mechanisms are almost non-existent or watered down, to the extent that proper public hearing processes and environmental and social impact assessment procedures are not seen as a necessary condition. This has been the story in other parts of Arunachal Pradesh as well, but the anti-dam protests by Buddhist monks in the streets of Tawang have greater national and international visibility.

The Dalai Lama's visit to Arunachal Pradesh, and particularly Tawang, has been routinely objected with official statements by China, and in 2017, during a visit to Tawang for some religious duties, he was invited as the Guest of Honour by the state government of Assam for its flagship river festival called the 'Namami Brahmaputra' River Festival.[20] This was a clear signalling by India to China, using the Dalai Lama's presence to the contested nature of riparian relations between India and China. The festival was also seen by many in Assam as a mark of imposed religious activities in the Brahmaputra River, primarily Hindu rituals of worshipping the river with lights and lamps, which are popular only to the Ganges River, and not naturally to the Brahmaputra. It was a conscious attempt of appropriating the cultural space of the Brahmaputra River to a 'Hindutva' political framing.

The presence of the Dalai Lama in the Namami Brahmaputra Festival was twofold, one was internationalizing the event, and making sure that China was paying attention, and second was of a symbol of token protest by India and the Dalai Lama together of the big hydropower dams being built by China in Tibet, on the same Brahmaputra River, known as Yarlung Zangbo in Tibetan and the Yarlung Tsangpo in Chinese. The irony here is that the sacred Buddhist spaces are being threatened by the hydropower dams being planned by India in Tawang, and you have the Dalai Lama sharing the stage with the Assam government to protest against the Chinese dams upstream in Tibet, and also reinforcing a Hindu 'sacredness' to the Brahmaputra River. The duality of creating Hindu sacred spaces on the Brahmaputra in Assam and the destruction of Buddhist sacred spaces in Tawang through dams are quite stark.

## Conclusion

The manifestations of such local community relations effectively look across the borderline, be it contested or otherwise, to take in both sides of the borderland,[21] where such human relations create, maintain, undermine and even evade borders, and challenge the idea of a national homeland which is sacrosanct and only determined and controlled by the nation state.[22] Another case in point is the opening of the Stilwell Road, which connects Ledo in Assam through Pangsau Pass in Arunachal Pradesh and northern Myanmar to Kunming in China's Yunnan Province, which can prove to be the pivot of India's Look East/Act East Policy. Local communities in Arunachal Pradesh such as the Tai Khamtis and the Singphos question the closed borders with northern Myanmar, with whom they have long-standing social and cultural connections, and want New Delhi to expedite the opening of the road, but are stuck in security concerns.

Another instance is the question of opening up of traditional trade and pilgrimage routes in Arunachal Pradesh such as Zemithang, which lies on the tri-junction of India, Bhutan and Tibetan Autonomous Region of China. The opening up of Zemithang has been discussed between the Bhutanese government and the Indian government in the past, for border trade and pilgrimage purposes, but has not materialized since. The question is of the imagination of the people of western Arunachal Pradesh of such closed border routes, and what the opening up of the same means to them in terms of physical, trade and larger spiritual connectivity with tribes of eastern Bhutan and Tibet. The people of Bhutan have to take a very circuitous and long route through South Bhutan and Assam to reach Tawang. The flow of pilgrims from Bhutan is because of the traditional relationships of Buddhist monasteries across the border with western Arunachal Pradesh, particularly evident during the visits of the Dalai Lama. This is the anthropological lens that India has to focus on in its borders, to ensure that infrastructure development is meaningful and sustainable for the border communities.

### Endnotes

1. Kashyap (2008).
2. Rahman (2019).
3. Verghese (2012).
4. Pandit (2009).
5. Chansoria (2011).
6. van Schendel (2013, p. 269).
7. Cited in Baruah (2008).
8. Scott (2011).
9. Verghese (1996).
10. Karlsson (2011).
11. Biswas and Suklabaidya (2008).
12. Rahman (2014).
13. Ibid.
14. Kashyap (2011).
15. Rahman (2014).
16. Aiyadurai (2011).
17. Rahman (2014).
18. Mitra (2013).
19. Gergan (2019).
20. Rahman (2017).
21. Evans (2013).
22. Diener and Hagen (2010):192 cited in van Schendel (2013).

## References

Aiyadurai, A. 2011. Wildlife Hunting and Conservation in Northeast India: A Need for an Interdisciplinary Understanding. *International Journal for Galliformes Conservation* 2: 61–73.

Baruah, S. 2008. Territoriality, Indigeneity and Rights in the Northeast India. *Economic and Political Weekly*, March 22.

Biswas, P., and C. Suklabaidya. 2008. *Ethnic Life-Worlds in Northeast India*. New Delhi: Sage.

Chansoria, M. 2011. China's Infrastructure Development in Tibet: Evaluating Trendlines. *Manekshaw Paper*, No. 32. New Delhi: Centre for Land Warfare Studies and KW Publishers.
Dzuvichu, L. 2013. Roads and the Raj: Politics of Road Building in Colonial Naga Hills. *Indian Economic Social History Review* 50 (4): 473–494.
Evans, R. 2013. The Perils of Being a Borderland People: On the Lhotshampas of Bhutan. In *Borderland Lives in Northern South Asia: Non-state Perspectives*, edited by Gellner D.N. Durham: Duke University Press.
Gergan, M.D. 2019. Geological Surprises: State Rationality and Himalayan Hydropower in India. *Roadsides* 1: 35–42.
Guha, S. 2008. China Protests Manmohan's Arunachal Visit. *DNA*, February 9.
Hazarika, S. 1995. *Strangers in the Mist: Tales of War and Peace from India's Northeast*. New Delhi: Penguin.
Jacob, J. 2015. Arunachal Pradesh in the Sino-Indian Boundary Dispute: Constant Claims, Changing Politics. In *Voices from the Border: Response to Chinese Claim over Arunachal Pradesh*, edited by Gurudas Das, C. Joshua Thomas, and Nani Bath, 48–62. New Delhi: Pentagon Press.
Karlsson, B.G. 2011. *Unruly Hills: A Political Ecology of India's Northeast*. Oxford: Berghahn Books.
Kashyap, S.G. 2008. Arunachal, Our Land of the Rising Sun, Says PM in Itanagar, Eye on Beijing. *The Indian Express*, January 31.
Kashyap, S.G. 2011. 17 Killed in Helicopter Crash in Tawang, Arunachal Pradesh. *The Indian Express*, April 19.
Mitra, Naresh. 2013. With Anti-Dam Mantra on Lips, Tawang Monks Hit the Streets. *The Times of India*, March 7. http://timesofindia.indiatimes.com/city/guwahati/With-antidam-mantra-on-lips-Tawang-monks-hit-the-streets/articleshow/18842465.cms.
Pandit, R. 2009. China Way Ahead on Border Infrastructure. *The Economic Times*, January 19.
Rahman, M.Z. 2014. Territory, Tribes, Turbines: Local Community Perceptions and Responses to Infrastructure Development Along the Sino-Indian Border in Arunachal Pradesh. *Institute of Chinese Studies Occasional Paper Series*, No. 7, June 2014. Delhi, India: Institute of Chinese Studies. https://www.icsin.org/uploads/2015/04/12/dc44619f98243f09109da6867923a56a.pdf.
Rahman, M.Z. 2017. *Dalai Lama at the Namami Brahmaputra River Festival in Assam: Mixed Signals for India-China Relations*. Institute of Chinese Studies (ICS) Blog, April 3. Delhi, India: Institute of Chinese Studies.
Rahman, Mirza Zulfiqur. 2019. 'Pickled' Infrastructure and Connectivity: Locating Community Engagement in Northeast India's Infrastructural Transformation. *HEINRICH-BÖLL-STIFTUNG*, June 7. https://in.boell.org/en/2019/06/07/pickled-infrastructure-and-connectivity-locating-community-engagement-northeast-indias.
Scott, J. 2011. *The Art of Not Being Governed: An Anarchist History of Upland Southeast Asia*. Yale Agrarian Study Series. New Haven: Yale University Press.
van Schendel, W. 2013. Afterword: Making the Most of Sensitive Borders. In *Borderland Lives in Northern South Asia: Non-state Perspectives*, edited by Gellner D.N. Durham: Duke University Press.
Verghese, B.G. 1996. *India's Northeast Resurgent*. New Delhi: Konark Publishers.
Verghese, B.G. 2012. The War We Lost. *The Tehelka Magazine*, Vol. 9, Issue 41.

**Mirza Zulfiqur Rahman** is a Visiting Research Associate at the Institute of Chinese Studies in New Delhi. He holds a Ph.D. in Development Studies from the Department of Humanities and Social Sciences, Indian Institute of Technology, Guwahati, Assam, India. His areas of academic interests include research on Northeast India, mainly on issues relating to transboundary water sharing and hydropower dams, roads and connectivity infrastructures, conflict and insurgency, peace building, development politics, migration and cross-border exchanges.

# Chapter 9
# Hydropower Projects in Tawang: Concerns and Flood Proximity Estimation Using GIS Analysis

**Amit Mukherjee**

## Introduction

India is the seventh largest producer of hydropower with a potential estimated at 148,700 MW at 60% load factor. At present, India produces 44,594 MW or 13.5% of the total electricity generated in India through hydropower projects. The development of the north-eastern states in India includes the construction of power grids and lines for the region. Their power generation potential for domestic, regional and national capacity has an immense capacity to add to this burgeoning requirement. To tap this potential, the north-east region of India can add up to 58,971–63,257 MW that would amount to 42.5% of total national capacity, of which production from Arunachal Pradesh is about 50,328 MW. It is 33% of national capacity that is being generated in the state of Arunachal Pradesh.[1] Out of the total 140 MOUs for hydropower projects in Arunachal Pradesh, about 13 have been planned in the Tawang River Basin. These projects in remote locations like Tawang without any precedent of major infrastructure projects caused some concerns among the local population in the region. These concerns gained prominence when protests against the developmental plans of hydropower projects became violent. The chapter is an attempt to support the comprehensive ethnographic study carried out for the project, but it has limitations due to the constraints of resources and methodology. It is a preliminary level assessment that has vast scope to be augmented with detailed study.

The agitation by the local NGO Save Mon Region Federation[2] in 2011 saw several studies being carried out to verify the claims and counterclaims towards the objections and clearances of hydel - power projects in the Tawang region. The agitation was carried out as local protests against an unhindered construction spree of hydropower

*Note*. All of Authors GIS outputs presented in the chapter are for illustration purposes only and not to scale.

A. Mukherjee (✉)
National Institute of Advanced Studies, Bengaluru, India
e-mail: toamitmukherjee@gmail.com



M. Mayilvaganan et al. (eds.), *Tawang, Monpas and Tibetan Buddhism in Transition*,
https://doi.org/10.1007/978-981-15-4346-3_9

projects in the region.[3] The objections raised included danger to flora and fauna and to endangered species like the red panda, black-necked crane, submersion of holy shrines and damage to the environment. Apart from these reasons, landslides, submergence of land and flooding due to the dam breach have formed the root of the concern. The decision to stop all projects was made based on the orders of the National Green Tribunal on the Nyamjang Chu and Tawang Chu (river), respectively. To verify the extent and the validity of some of the concerns raised against the construction of the power project, an objective-based assessment has been attempted in this chapter which has been done using geospatial analysis through satellite imagery.

The environmental impact on the local habitat of various species of birds and primates, local ecology and livelihood of the region would get influenced due to the construction of large dams. The initial clearance received for the projects were revoked as a result of protests from the local community.

Since then, there have been several approaches by various organisations to persuade the local population about the benefits of hydroelectric projects. This mandates a comprehensive analysis of the region and the problems. It primarily includes an inquiry into the status of opinion held by the locals regarding the issue. The search for remedial solutions and possible conditions that would be agreeable to the local population for withdrawing their objections in the area needs examination. It is essential to recognise the aspirations of the people of the region and seek suggestions for the improvements in the area.

The reports and statements from several research institutes and government organisations are considered in this analysis. One of the areas of preliminary investigations was to find out the proximity of the towns and villages in Tawang to the river bed. To envisage possible scenarios of river overflow that might inundate the banks of the river causing landslides and subsequently flood-like situations due to river blockage. Further, computational models using Geographic Information System and Satellite Imagery were carried out for the analysis. Some of the results of the GIS-based analysis have been discussed in this chapter.

The study area of 910 41′ 25.21″ E long. and 270 45′ 51.28″ N lat. that includes the Tawang River Basin and the Nyamjang Chu River Basin is considered (Fig. 9.1).

## Methodology of the Analysis

The use of Geographic Information System (GIS) and satellite imagery has been the tool for carrying out this analysis. Topo map, satellite data and tools of GIS were used for analysis. Spatial data attributes and non-spatial data of geographical features were synchronised. Sentinel and Landsat data, DEM data and open-source maps were used to build, verify and validate some of the points that have been taken into consideration. The secondary sources along with field data and personal interviews are also used. As part of the analysis, the thematic maps were generated (presented here as simulated/pictorial, not-to-scale images) that include features like general drainage, catchment and sub-watersheds, relief map, in addition to river gradient

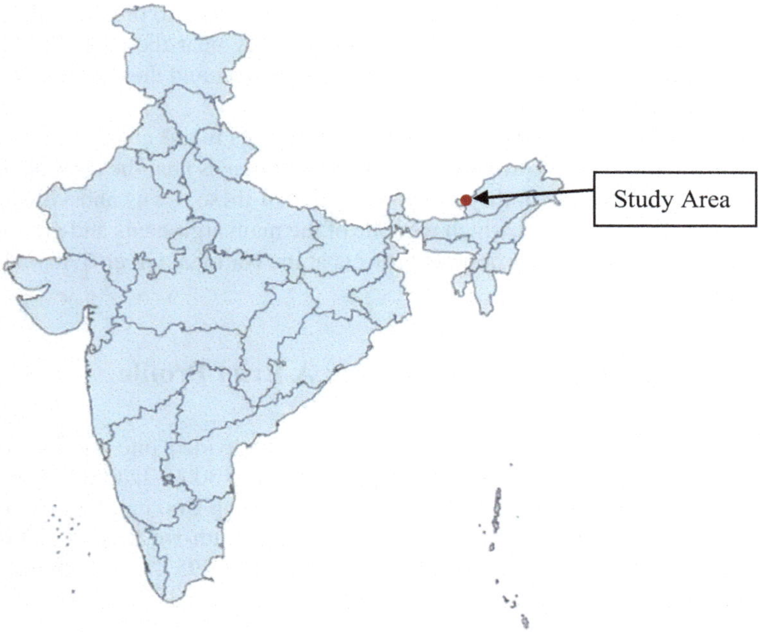


**Fig. 9.1** Map of the study area, Google Earth Imagery. *Source* http://www.mea.gov.in/

**Fig. 9.2** Map overview of Arunachal Pradesh (not to scale). *Source* d-maps.com

profile of the Tawang Chu and Nyamjang Chu. This was done to illustrate altitudinal difference between towns and villages and river valley near the proposed barrage site. The relative altitude and slope were plotted to understand the degree of erosion or more importantly submergence.

Thus, the objectives of most of the geospatial analysis in this chapter are twofold: firstly, to estimate the flood proximity of towns and villages near the river banks and secondly, to check for possibility of submergence of these towns and villages and religious sites along the river, which was one of the main arguments and grounds for the appeals against mega-hydropower projects at the National Green Tribunal.

## Topography and Rainfall in Tawang: A Brief Profile

Tawang district has an area of about 2085 $km^2$, subdivided into subdivisions of Lumla, Jung and Tawang. It is situated at an elevation of 3000–22,000 ft. It has two river streams that include Tawang Chu and the Nyamjang Chu entering from Tibet at the north. The confluence of the rivers occurs at the south-western tip of Tawang, approximately at the coordinates of 91.414.231 E and 27.2957.751 N, about 30 km south of Indo-China line of actual control and 5 km east of Indo-Bhutan border. The area of Nyamjang Chu and Tawang Chu basins is displayed in Fig. 9.3.

Initial analysis was carried out to identify the suitable extent of project development between Zemithang and Kumba villages. River bed elevations at Zemithang

Fig. 9.3 District map of Arunachal Pradesh, India. *Source* tawang.nic.in

and at the confluence of Nyamjang Chu with Tawang Chu near Kumba village are around elevation 2106.2 m and elevation 1041.4 m, respectively. The topographical details including physical features, villages, religious monuments and other structures falling within the reach from Zemithang to the confluence were identified to assess possible impacts of placing the project structures in development alternatives.[4]

At the barrage sites, the terrain of the rivers is flat and has loamy soil. Given its characteristics, loamy soil tends to erode if it is found at higher slopes along the river run-off areas.[5] The Nyamjang Chu river bed area is 287 $km^2$ of which only 1.74% is highly erodible vulnerable zone, 22.54% is from moderate to low, and 9.08% is of low vulnerability. The high risk area falls under 0.04%, moderately high is 4.05%, moderate range of 34.59%, moderately low was mapped at 51.95% and a low of 9.37%.[6] Given low indexing of erosion and landslide vulnerability zones, the concern for these hazards is also appears minimal under standard and general circumstances.[7]

As reported, the annual average rainfall measurement from three main gauge stations at Yusum and Tawang is 2053 mm. The annual rainfall in the region is between 1500–2000 mm and the annual rainfall in Tawang is 1600 mm. The weather and climatic conditions in the region are influenced by the topography and terrain of mountainous peaks covered with perennial snow. In addition, the region also experiences frequent floods and landslides usually caused by excessive rains and cloud burst.

## The Nyamjang Chu River Basin: Terrain Profile and Dam Concerns

The Nyamjang Chu River Basin lies in the north-west area of Arunachal Pradesh with its catchment spreading across the international border covering part of Tibet. Nyamjang Chu originates from snow-clad peaks in Tibet at an elevation of 6400 m approximately and flows in India from north to south up to its confluence point with Tawang Chu before entering Bhutan, having run a course of about 40 km inside India. It is a tributary of the River Manas.[8] The total catchment area of the Nyamjang Chu up to the confluence with Tawang Chu is about 3170 km. The catchment area intercepted up to the diversion site near Zemithang is about 2650 km. The catchment area is mostly of tropical wet climate and supports dense mixed forest. The area is characterised by hills with steep gorges and deep rugged valleys with streams feeding the Nyamjang Chu river system of which Takhsang Chu and Sumta Chu are major contributors.[9] The construction of low-height construction barrage, surface de-silting basins, head race tunnel, pressure shaft, underground powerhouse and tail race tunnel for generation of powerhouse from six Pelton turbines were planned.[10]

The Nyamjang Chu river confluences with the Tawang Chu, before flowing into Bhutan, where it is called Gamri Chu[11]. The stream flows take sharp bends and meet

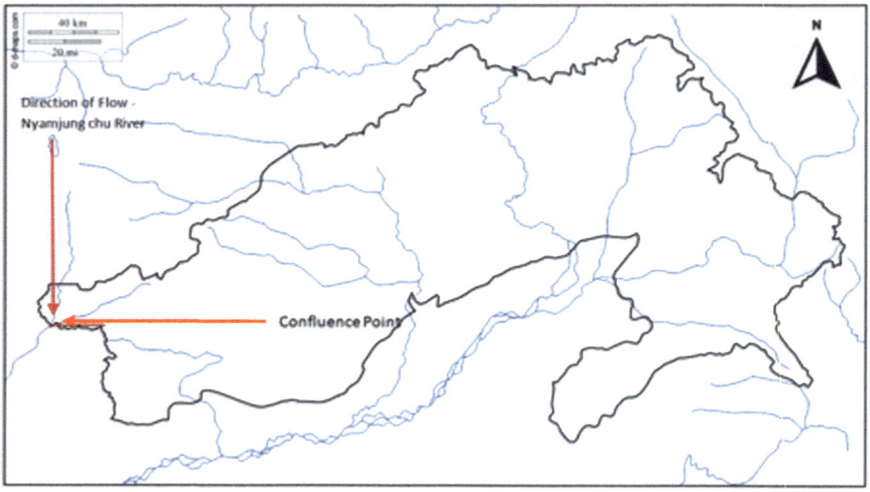


**Fig. 9.4** Nyamjang Chu river flow north–south direction. *Source* d-maps.com

with the main river. Along with this, there are a number of first-, second- and third-order streams joining the main river. The number of first-order streams on the right bank is much more than the left bank. There are several first-order streams on the right bank of the river to which several second combine to fill third-order streams (Fig. 9.4).

## *Hydropower Projects in Tawang (Proposed/Under Construction)*

The Nyamjang Chu River Hydropower Project is a run-off-the-river scheme with a capacity to generate 780 MW with an underground powerhouse plan.[12] It will harness the hydropower potential of the Nyamjang Chu River with a diversion barrage near the Zemithang village and a powerhouse near the confluence of the Tawang Chu with the Nyamjang Chu (Fig. 9.4). The total potential of the basin is around 900 MW. The project proposes to generate 780 MW in an underground powerhouse. It will consist of 11 m high barrage, 175 meters in length and a tunnel of 23 km approximately, thereby diverting the water to the powerhouse downstream near Lumla.

The Government of Arunachal Pradesh awarded the work of development of the hydropower potential in the Nyamjang Chu River Valley (also known as the Pangchen Valley) in Tawang district to Bhilwara Energy Limited (BEL) in 2006. WAPCOS Ltd. and RS Envirolink Technologies Private Ltd were commissioned to carry out the EIA and EMP for the proposed project. They undertook the EIA studies in 2007–2008. The primary data that was collected for the EIA was initially

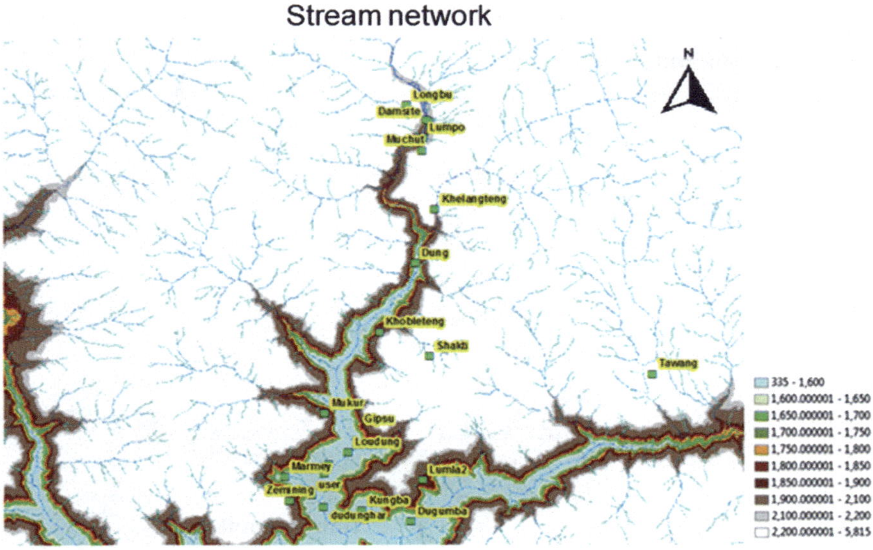


**Fig. 9.5** Stream network of Nyamjang Chu River (illustration purposes only). *Source* Author's GIS

done by WAPCOS for three seasons (August–September 2007, December–January 2008, April–May 2008) and then was later given to another agency (RS Envirolink Technologies Private Limited) for three seasons (April–May 2008, July–August 2008 and November–December 2008).[13] There is also a proposal to build a tunnel at the dam sites on the Nyamjang Chu River to divert excess water in case of an increase in water volume behind the dam. The proposed site is expected to take advantage of a drop in altitude of about 800 m from the dam to the exit-site (Figs. 9.5 and 9.6).

Several towns and villages exist along the side of the Nyamjang Chu River region; however, for the purpose of the analysis, a total of 17 neighbouring towns and villages were taken. These locations were chosen based on the lowest altitude along the river, using publicly available data. Starting at the northern most position is, Longbu (2578 m), Zemithang (2134 m), Lumpo (2058 m), Muchut (2265 m), Khelangteng (2223 m), Dung (1621 m), Khobleteng (2017 m), Shakti (2682 m), Gispu (1996 m), Mukur (2011 m), Loudung (1224 m), Dormeleng (1195 m), Marmey (1530 m), Zemining (2057 m), Dudunghar (1167 m), Kungba (1400 m), Lumla (1947 m) and Dugumba (1420 m).

The terrain and satellite data and other information for analysis of Nyamjang Chu flood inundation have also been used. Visual inspection of recent satellite data of the river valley indicates no large signature of flooding at the river terraces. A natural land formation on the river bed near the dam site creates a decrease in width at the location that are shown after profile section 7 (Fig. 9.9). Beyond this point, the flow of the river downstream is narrow in width confined to a deep valley.

The villages existing on both sides of the river and river alignment near dam site are shown in Fig. 9.9. The Nyamjang Chu River flowing from north to south can be

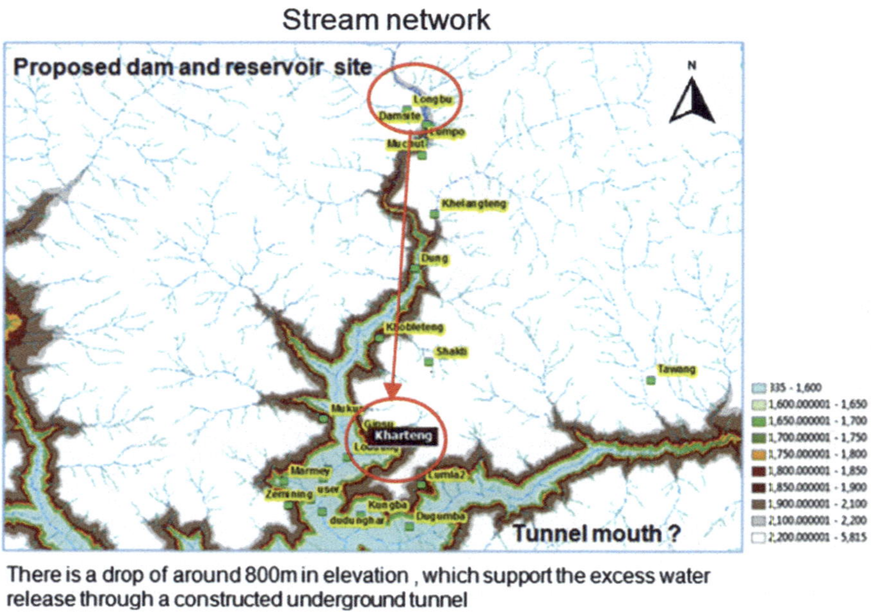


**Fig. 9.6** Stream location for dam site and tunnel exit—Nyamjang Chu (illustration purposes only) (not to scale). *Source* Author's GIS

observed. Contours were plotted at an interval of 20 m. The barren area and potential landslide zones have also been marked. The heights of the towns and villages have been compared with the altitude of river valley level using digital elevation model. However, the landslide zones that have formed over time can add to the increase in water level up to a certain extent. There are newer landslide zones that are forming due to heavy deforestation that has occurred in the region over the recent years. The soil formation, gradient, rainfall quantity, deforestation, rock and crust formation are responsible for the fragile environment of the region. The flooding and other natural calamities would not be triggered merely due to dam breach and construction of hydropower projects. The increase in flooding is also restricted due to the presence of steep slope in the river valley along the flow direction of the river.

## Submergence Concerns of Religious Site

*Gorsham Chorten*: The concerns of submergence of this sacred Buddhist site in the location have been cited among others, as an important reason for opposition to the dam projects. The visual interpretation and comparison of altitude of river bed, river water flow and the base height of the Chorten provide a fair idea of the possible submergence scenario (refer Picture 9.1 and Figs. 9.7 and 9.8).

**Picture 9.1** Gorsham Chorten. *Source* tawang.nic.in

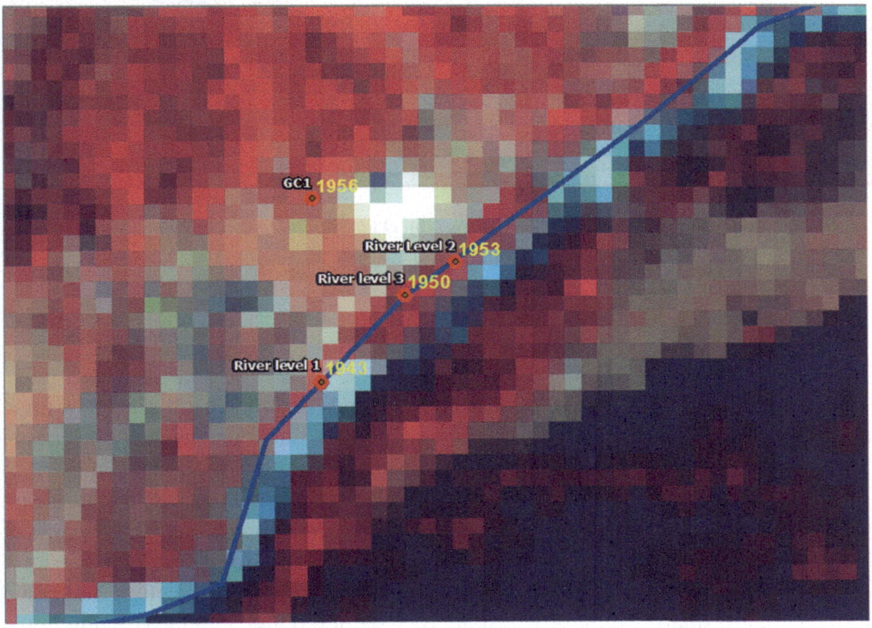


**Fig. 9.7** Site altitude of *Gorsham Chorten* and random river bed locations (illustration purposes only). *Source* Author's GIS

The base height at location of *Gorsham Chorten* in the image is 1956 m, whereas the three random points that were selected in its proximity are between 1945 and 1953 m. The difference in altitude of about 10 m at the lowest points makes the site susceptible to flood-like situation under heavy flooding or blockage of run-off of the river flow downstream. These issues can however be overcome by constructing appropriate embankments and planned diversions.

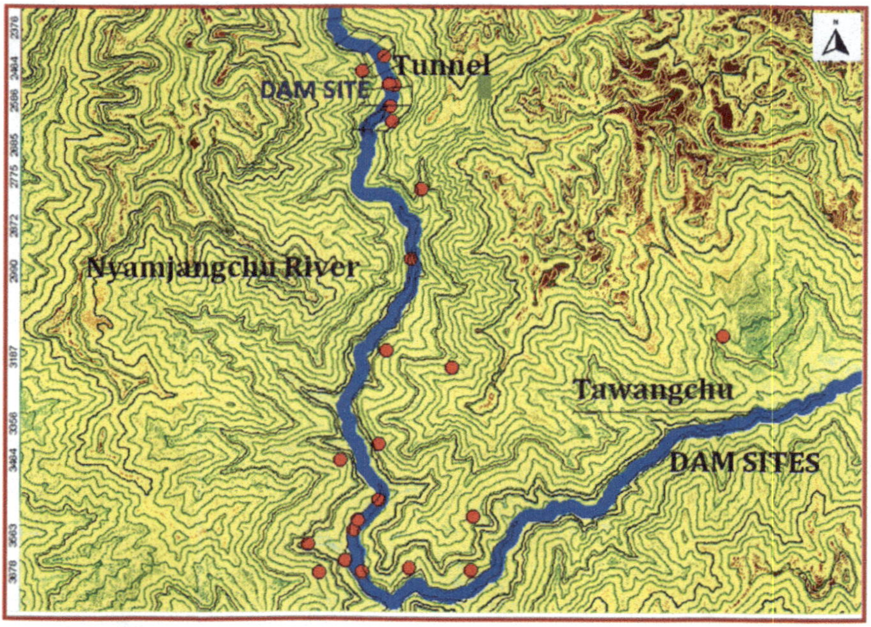


**Fig. 9.8** Contour profile map of River Nyamjang Chu Dam Site and river area (illustration purposes only)

## Cross-Sectional Profile Analysis

The cross-profile map along the river valley indicates that the best reservoir volume and construction parameters are at a given location. It illustrates the cross-sectional profile of the river before the barrage site on the Nyamjang Chu (river) (Fig. 9.6).

Contour map draped over colour-coded DEM indicates changes in the terrain elevation and ascertains the effect of river flow on neighbouring areas. The Nyamjang Chu (river) takes an S bend in its flow from the north eastern side to the south westerly direction. The potential reservoir area starts to broaden after cross-section 2 and then more so after 4, finally being broadest at cross section 6 and 7 and narrowing down at profile number 8 (Fig. 9.9). The narrow profile after section 7 makes it suited for construction site.

The image (Fig. 9.9) displays cross section in the river flow and its profile on the right side of the image. The most stable profile is seen after section 7, which is an ideal location cited for building of the barrage. The contour interval in the map is 20 m; however elevation change notation at 50 m interval was used in the map.

## Colour-Coded DEM

The villages and towns near the dam site, namely Zemithang, Lumpo and Socksten, are less likely to be flooded due to dam breach or release of water because of the slope in the region. The run direction of the river and proposed construction of a pipeline for regulated water discharge would add to the reduction in any flooding activity. The height of these locations is at 150–200 m from the base of the river valley as seen on the map. This indicates that flood due to river overflows, dam breach and dam reservoir storage would not create the problem. (The above observation runs in contrast to the notion of several local people in Zemithang Valley and respondents interviewed in Tawang.) However, it must be noted that any grazing area for cattle and all types of present-day land use land cover that are at the banks and up to 10–15 m average altitude may get submerged near the barrage where reservoir storage is being built. In Picture 9.2, low-level sheds and cottages are too close to the proposed dam discharge area. Thus, the level of water due to the dam discharges may lead to the submergence of such regions.

All three towns marked in cross-profile section (Fig. 9.12) indicate that they are higher than the river bed (average altitude 2115 m). The height difference at profile A is 30 m, in profile B it is about 180 m, and in profile C it is 40 m, while at D it is more than 350 m higher than the river bed. The towns at locations A and C are

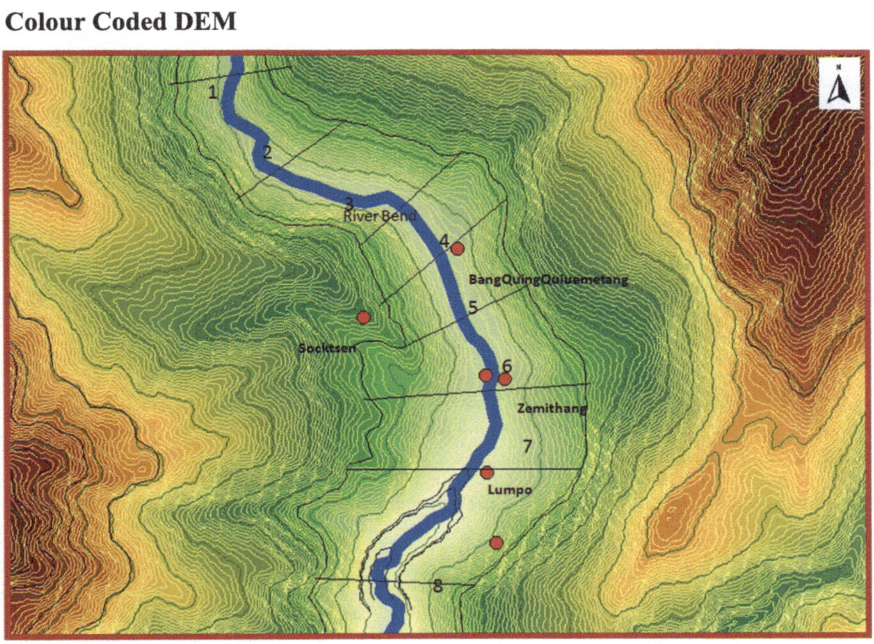


**Fig. 9.9** Nyamjang Chu cross-sectional profile area on a contour map ~1.5 km (illustration purposes only)

**Fig. 9.10** Nyamjang Chu cross section of profile 6. (Illustration Purposes only) *Source* Author

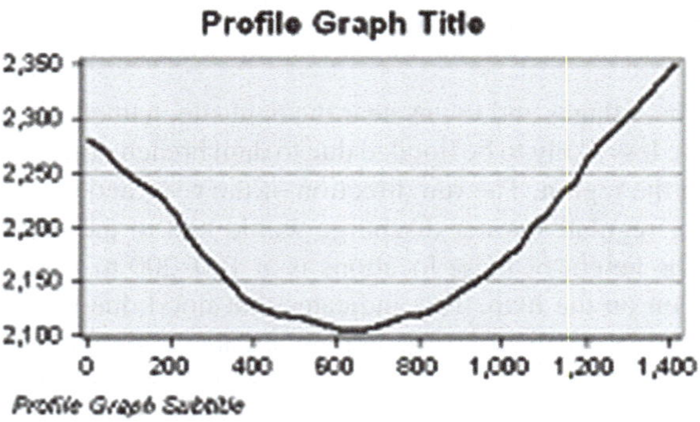


**Fig. 9.11** Nyamjang Chu cross section of profile 7. *Source* Author

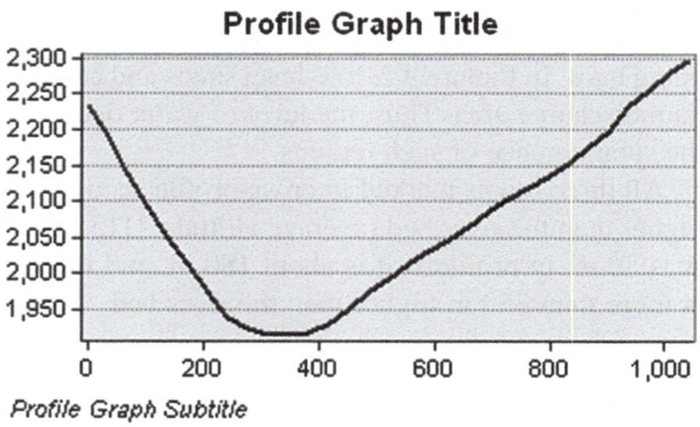

**Picture 9.2** Nyamjang Chu—altitudinal difference and visual inference

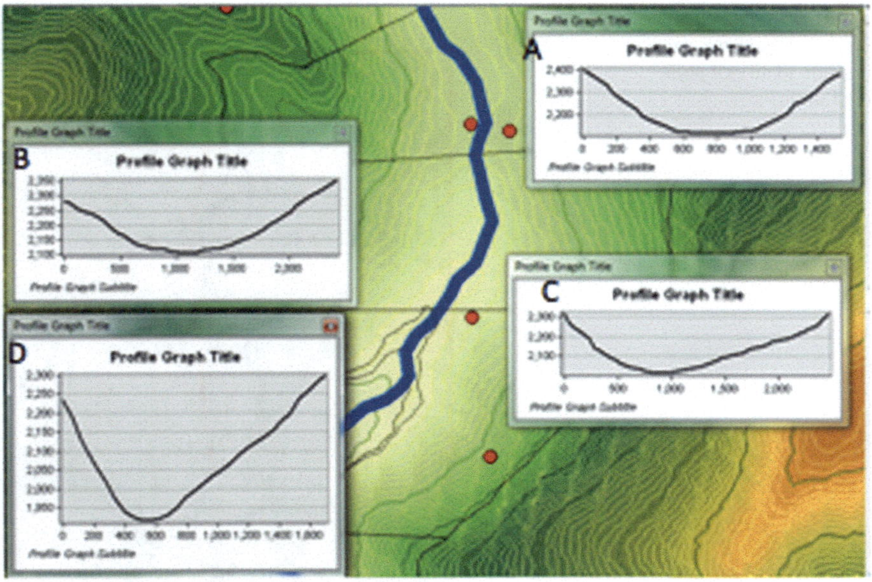


**Fig. 9.12** Simulated representation: river flow profile north to south—villages near bank of river

at their extremities and maybe effected due to a sudden rise in water owing to few exceptional circumstances like cloud burst, glacial meltdown, etc. These places are parts of Zemithang, Lumpo and Socksten that are closest to the river bank. However, the majority of the inhabited locations appear to be safe from river overflow or a dam discharge. This has even been noted in earlier published reports.

The location cited for the dam site is best suited within the areas under planning, for the construction because of its natural topographical conditions. Since the proposed site is of $(6 \times 130)$ 780 MW total capacity, building a 100 MW capacity[14] hydroelectric power for trials and later on changing the production capacity would create redundancy and add excess construction costs. The re-engineering process would also add to the challenges, thus making the enterprise redundant and futile. However, dam construction should be done after taking into consideration all the concerns and views that created the stopping of the activity once Green Tribunal and Court order clearances have been attained.

As per media reports in 2015, the flooding in the Zemithang region induced by huge rainfall in the upstream and catchment area disconnected the region from the rest of the district of Tawang and other parts of Arunachal Pradesh. The drainage of this excess water floods the Nyamjang Chu River.[15]

Building plans of a diversion for the river can assist in prevention of river overflow in the downstream direction due to any possible flooding. The construction of an underground pipe system that is proposed is expected to provide such a flow control (Fig. 9.13).

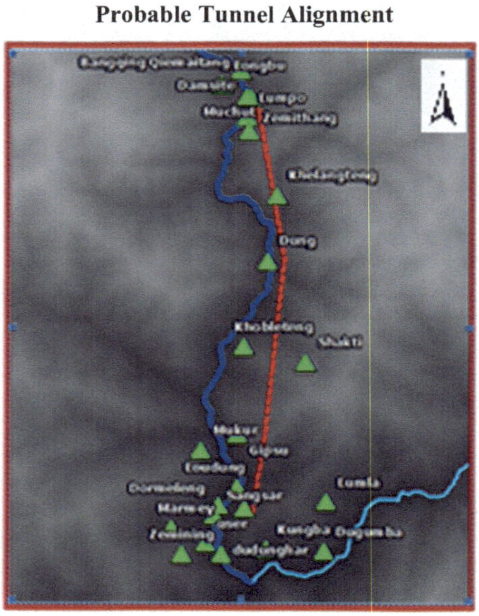


**Fig. 9.13** Simulated image—tunnel from dam site diversion to an exit point near Lumla

## The Tawang Chu (River) Basin: Terrain Profile and Dam Concerns

The Tawang Chu (river) is located in the north-east of the district of Tawang and flows in a north to south direction. The basin area has 75% mountainous region of higher Himalayas covered with steep terrain and rocky and precipitous slope. The basin is said to be susceptible to climatic changes, because of its 'geo-ecological fragility'.[16] The height of Tawang Chu (river) is around 6000 to 22,000 ft, and most inhabitants are found at lower altitudes.[17] The area for analysis is mainly of environmental and natural hazards like flooding in towns and villages on the embankment or on the order of the river's run-off direction from north-east to south-east. The base altitude of river bed from near towns in this region was estimated using open-source altitude maps, as follows: Jang (2268 m), Jangda (2168 m), Shyro (1976 m), Gomkelleng (1942 m), Kudung (1592 m) and Bumteng (1642 m). Most of the towns were found to be at a well-located height that would be able to avoid any damage from river flooding due to a dam breach.

The Tawang hydroelectric projects are described above and depicted in Fig. 9.14, and the towns and villages are near the banks of the river. The heights have been derived from the DEM at 30 m resolution (Figs. 9.9, 9.13, 9.14).

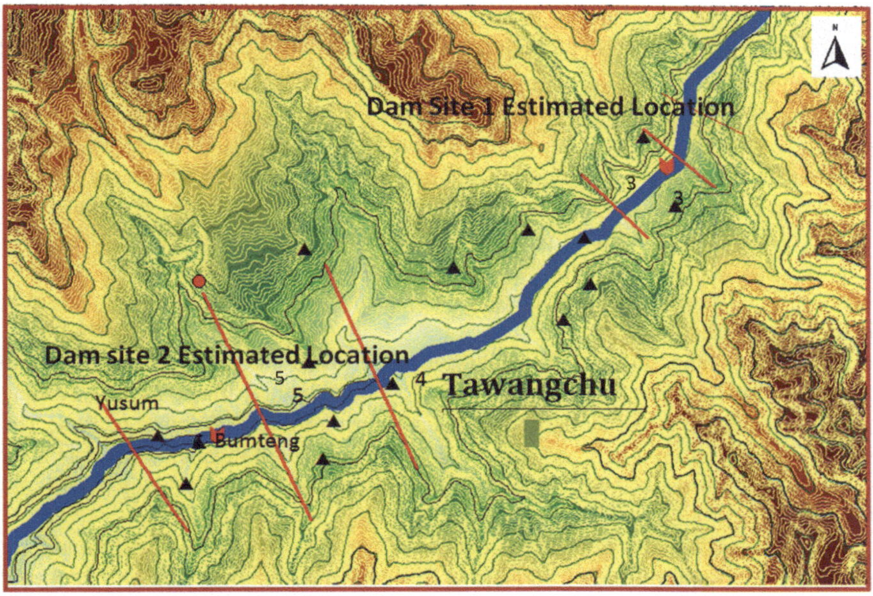


**Fig. 9.14** Contour profile map of River Tawang Chu Dam Site and river area (illustration purposes only)

## *The Tawang Chu Hydropower Projects*

'The Tawang Hydroelectric Project is proposed on Tawang Chu (river) and is located at downstream of fore-bridge near Jang (2.5 km downstream of Nuranang Chu powerhouse). The dam site is located about 28 km from Tawang Town on Tawang—Jang road and powerhouse site are located near Yusum village.[18] This dam will have an installed capacity of 600 MW (3 × 200) with 2963.00 MU (in a 90% dependable scenario), costing 6131.69 crores (May 2016 price level), with a barrage height of 26 m, 130 cm long; Head Race Tunnel 1.7 m dia, 13.987 km long; horseshoe shaped, with a surge shaft of 18 m dia, 125 m high orifice type; and a vertical Francis turbine 1680 m long, 7 m dia horseshoe.'[19]

The Tawang II project design is for constructing an 800 MW power project with the design energy of 3622.00 MU (in a 90% dependable year for discharge), on Tawang Chu (river), at a reported cost of approximately INR 78,298 crore on September 2015. The technical design features include diversion structure at 28 m height of the barrage with RCC raft piers. The head-shaped tunnel dimensions are 1 no., 8.1 m dia, 15.989 km-long horseshoe shaped; the surge shaft is 1 in no., 20 m diameter, 131 m high; the powerhouse will have underground Francis vertical axis turbine; the pressure shaft is from EL 1461.50 m to 1296.00 m with underground steel lined 5 m diameter and circular shaped. The tail race tunnel is supposed to be 8.1 m dia and 170-m-long horseshoe shaped.[20]

Both locations form a width pattern seemingly suited for construction of the barrage and diversion. It can be seen from the figure that elevations of the river level are much lower than the adjacent towns and villages. Given that the highest rise in water levels is generally not more than 50 m, the possibility of dam-led floods due to excess water release is unlikely excepting in locations like Bumteng.[21] As is apparent, in most cases the height difference between the river levels and town in this region is substantially high and hence chances of flooding appear unlikely.

The cross-sectional profile map analysis across the Tawang Chu (river) valley indicates that the best reservoir volume and construction parameters are at a site location. The following image displays cross section in the river flow and its profile on the sides of the image.

The contour map draped over colour-coded DEM indicates changes in the terrain elevation and ascertains the effect of a river flow on neighbouring areas. The Tawang Chu (river) takes a curve along its North East to South West drection of flow. The potential reservoir area is broad before cross section 2 and then subsequently after 5, finally being broadest at cross-sectional profile numbers 5 and 6 (Figs. 9.15 and 9.16).

The villages and towns near the dam site, namely Jung, Shyro and Gonghkar near cross-sectional profiles 1 and 2 and Kudung, Bumteng and Yusum near cross section 5 and 6, are the areas of primary concern for flooding. However, height of these locations is at least 15–30 m at the lowest level at Shyro and Bumteng as seen in the map (Fig. 9.14). The steep drop in height will prevent water accumulation and flooding under general circumstances. A general phenomenon of a narrow profile considered a suitable site for water collection using a barrage/dam. A broad base would increase spread of the volume of water reducing pressure, whereas narrow profile increases pressure raising the water level in the dam.

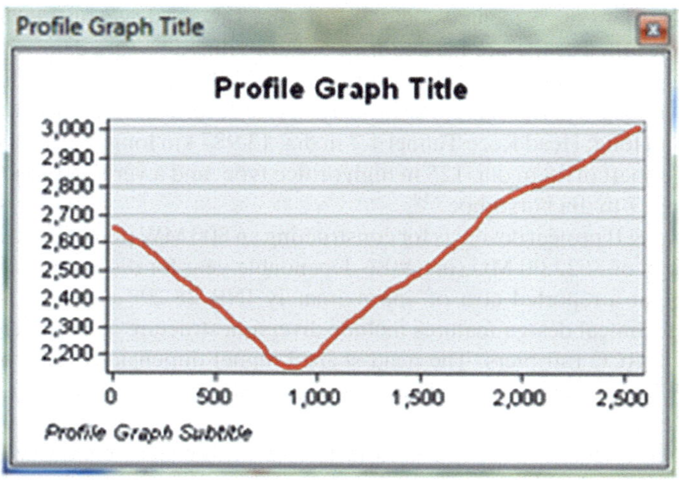


**Fig. 9.15** Cross section profle 5

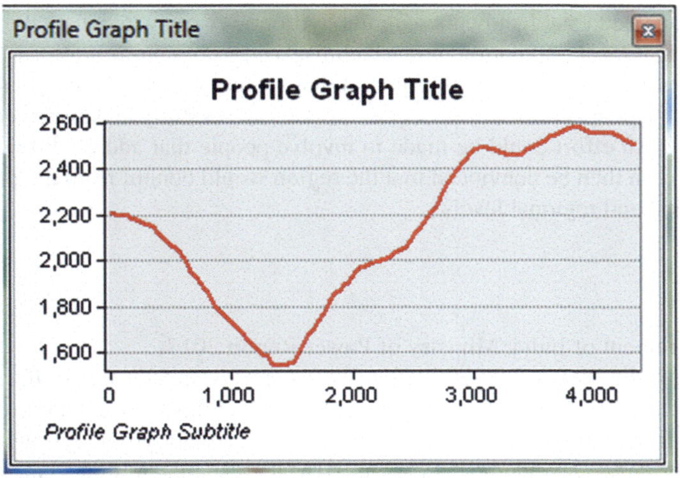


**Fig. 9.16** Cross section profle 6

## Conclusion

A preliminary estimation of the Nyamjang Chu and Tawang Chu hydropower projects individually leads to suggest that most of the dam construction sites do not have a susceptibility to flooding and submergence of any major habitat regions, under general conditions. This is mainly due to altitude difference, slope of the river's run-off terrain topography, provisions for drainage through tunnels and a strong dam's construction that can restrict flooding caused in the region even if there is rainfall by storage and drainage.

Several locations where flooding could be a concern due to blockage of dam discharge or due to landslide that may trigger the dam discharge to flood in a particular region. This can be prevented by additional precautions like building proper preventive embankment and diversions. The gap between scientific expert knowledge and public perception could be observed when it comes to gauging the popular sentiment for the mega-hydropower project. Thus, public awareness programmes would be a viable idea for reducing the fear among local population that the construction of mega-project dam is detrimental to their interest in general and concerns of catastrophe due to incidents like a dam breach or blockade of flow. Furthermore, a detailed and complex analysis of flood modelling done under various simulated circumstances could and should be carried out to demonstrate the likelihood scenarios to both ascertain and demonstrate safety of the locations of the project site.

Hence, the lack of confidence of the local population, viz. the dam and hydroelectric projects, could be overcome by adopting an alternative approach by the government agencies. Confidence building measures to improve public perception and understanding to mitigate the communities' concerns is an important step towards reconciliation of the issues.

The concerned establishments with the hydropower projects must take into consideration that local population may have misgivings or genuine concerns that need to be addressed. A case study of similar agitation that was resolved through dialog and taking the people into confidence has worked successfully in dam projects across the country. An effort could be made to involve people that add credibility to such works who can then be convinced that the region would benefit from the projects, at both national and regional levels.

## Endnotes

1. Government of India, Ministry of Power Report 2017.
2. Joan Martinez-Alier, *Buddhist monk killed in resistance to dam project in India*, May 5, 2016, http://www.ejolt.org/2016/05/buddhist-monk-killed-resistance-dam-project-india/.
3. Border Area Development Program (BADP), Annual Action Plan, Annexure II, Arunachal Pradesh, June 11, 2016, http://www.arunachalplan.gov.in/html/docs/badp/BADP_Annual_action_Plan_2016-17.pdf.
4. Environment Impact Assessment, 29 January 2010, http://www.moef.nic.in/sites/default/files/EIA-NJC.pdf.
5. Manual, https://www.michigan.gov/documents/deq/wrd-cswo-sesc-manual-unit7_556419_7.pdf.
6. Ibid, p. 4.
7. S.K. Barik, *Perspective plan for development of Tawang basin*, Cumulative Impact assessment of Proposed Hydel Power Projects, Determination of Basin Carrying Capacity, Landscape Level Biodiversity, Management Plan, June 2015. Sec. 3.3.11.
8. WS Rocky, G Rongmei, SC Kesari, Senior Geologists Engineering Shillong Division, North Region Shillong, *Report on the preliminary post disaster landslide studies in and around Tawang town district*, Arunachal Pradesh, Geological Survey of India, 2016–2017. p. 4–8.
9. Manual, https://www.michigan.gov/documents/deq/wrd-cswo-sesc-manual-unit7_556419_7.pdf.
10. Ibid.
11. WS Rocky et al, 2017.
12. Survey of the waters of Bhutan and Fisheries Potential - fao.org
13. Ibid.
14. NHPL NJC Power Limited, Hjhydropower.com/projects, NJCHEP.
15. *A critique of Nyamjang chu Hydro-Electric Power Project Environmental Impact Assessment EIA*, *Environmental Management Plan (EMP)*, ResearchGate Publication. http://njchydropower.com/projects.html.
16. Environment Impact Assessment, 29 January 2010, http://www.moef.nic.in/sites/default/files/EIA-NJC.pdf.
17. Environment Impact Assessment, 29 January 2010, http://www.moef.nic.in/sites/default/files/EIA-NJC.pdf.

18. OJ Ranjan, S Anand, BW Pandey, *Cultivation Ecology in Tawang Chu (river) Basin, Arunachal Pradesh*, Delhi School of Economics, University of Delhi.
19. NHPL NJC Power Limited, Hjhydropower.com/projects, NJCHEP.
20. Ibid.
21. Ibid.
22. *Flood situation grim in Arunachal Pradesh*, www.tribuneindia.com/news/nation/flood-sitation-grim-in-arunachal-pradesh/126872.html.

**Amit Mukherjee** is an Assistant Professor in the International Strategic and Security Studies Program at the School of Conflict and Security Studies in National Institute of Advanced Studies (NIAS), Bangalore. His specialization is in Military GIS and application of GIS for traditional and non-traditional areas of National and International Security. His research interests include homeland security, weapons and warfare, military affairs and strategy and border security among others. He holds Post-Doctoral Fellow at the National Security Studies Center, University of Haifa, Israel, from 2013 to 2016 with an Israel Asia Leaders Fellowship in 2014 and a visiting fellow at Institute of Defence Studies and Analyses (IDSA), New Delhi, 2017.

# Chapter 10
# Locked in Hydropolitics: Understanding the Local Protests and Differing Dynamics in Tawang

**Sourina Bej**

## Introduction

On 2 May 2016, the western border district of Arunachal Pradesh, Tawang, became the centre stage of anti-dam protests that took a violent turn when two protesters lost their lives in police firing with 10 more seriously injured. By mid-afternoon that day, prohibitory orders under Section 144 were imposed in the Tawang town and the army was called to stage a flag march.[1] This was the first time violence in the form of police firing took place in the district and also among the Monpa tribal group. Since few years, the Monpas in the Tawang district have been protesting to stop the mega-hydropower projects on the rivers of Tawang Chu and Nyamjang Chu but the 2 May incident unfolded a local resistance that was slowly being mobilised in this strategically important border district of India. The social resistance has not only brought the Monpa tribal group together to stand against the state's infrastructural development policies, but following the firing, their plea[2] to stop the mega-dam project became the first petition to be accepted by the National Green Tribunal (and not challenged by the Ministry of Environment, Forest and Climate Change). After the petition, the 780 MW Nyamjang Chu Project stood cancelled. This is not the first time that dam projects have also been contested, but it is one of a situation that it was defeated in India.

Starting with the *Narmada Bachao Andolan*, the anti-dam social movements have been purposeful, organised and striving to work towards a common goal: the government's attempt to capitalise on the river to generate power and electricity has to take into account the voice of the river-dependent (riverine) communities that would be disrupted by the dam construction. Thus, these protests have become a method of mass mobilisation to generate a political voice for the under-represented and bring

S. Bej (✉)
International Strategic and Security Studies Programme, National Institute of Advanced Studies, Bengaluru 560012, India
e-mail: sourinabej92@gmail.com



M. Mayilvaganan et al. (eds.), *Tawang, Monpas and Tibetan Buddhism in Transition*,
https://doi.org/10.1007/978-981-15-4346-3_10

forth the dysfunction between systems that continues to widen without mitigation in sight.

In terms of the common goal, the protest by the Monpas is not different from the discourse on the politics of dam in India. At the same time, the resistance is unique when it comes to the narrative, underlying causes and the tools used to legitimise their voice. The paper aims to look at the rhetoric that the anti-dam protests by the Monpa make and how different is it from similar local resistances occurring in the northeast region. In addition, much research has been done to understand the interstate politics of dams and the intra-state politics of displacement and resistance to dam construction.[3] But little focus has been given to understand a local protest that is born out of the grass-roots systemic mismanagement of resources but goes to acquire an international or transnational dynamic owing to the larger geography and location of the dam, the significance of the river and the site of dam construction (in this case the Tawang district and the northeast region of India). This chapter tries to comprehend this overlapping of external dynamics into the internal causes of the protests and locates the answer in the national-level decision-making.

## The Development Drive in Northeast India: A Perspective on Tawang

It is imperative to understand the local agitation against the dam not in isolation of the national and the sub-national decision-making processes. Also, it is vital to understand as to why the push for dam construction became necessary in the region as it not only underlines the causes in the inter-tribal and the intra-state resentments but also places the agitation in the larger rationale of the central government's policy to develop. This scope to development served a dual purpose: it not only integrated the nationwide development process but through the narrow lens of modernisation and mechanisation (dam-building) also turned the northeast rivers into the powerhouse of India. As India looked to improve its energy deficit, the supply chain had to be increased and this was possible only by shifting to the generation of renewable energy with less carbon footprint.

According to the estimates of the Central Electricity Authority, the Northeast India's potential to generate around 58,971 MW of hydropower[4] was identified more than a decade back. In this, Arunachal Pradesh alone had a potential to generate 50,328 MW of energy—the highest in the country.[5] But until 2000, this knowledge was not turned into plausible actions and the region remained out of any pro-development policy framework of the centre. Now, the expectation and realisation among different sub-national political elites that their state could make windfall gains from hydropower have facilitated New Delhi's drive to put hydropower development on the eastern Himalayan rivers on a fast track.[6]

Thus, in May 2003, when Prime Minister Atal Bihari Vajpayee launched the 50,000 MW Programme, it was seen as a policy course correction by the centre to bring the desired development and the infrastructure to the region. According to the Vajpayee Initiative, 50,000 MW of installed capacity of hydropower was identified to be constructed across 16 states in the country. The pivot to this scheme was Arunachal Pradesh (apart from Manipur, Assam and Sikkim) where the hydropower potential was largely untapped and of the 162 hydropower sites identified for the initiative, 42 was to be installed in Arunachal Pradesh.[7] By May 2008, in tune with the initiative, a flurry of memorandum of understanding for hydropower projects was signed by different states in Northeast India with the maximum being signed by the Arunachal Pradesh government. This led the former Environment Minister Jairam Ramesh to use the term "MoU Virus" to describe the speed with which each dam projects were signed (Mishra, 2014). As of 2017, Arunachal Pradesh alone has 50,328 MW of India's total 148,701 MW hydropower potential (as per the Central Electricity Authority) and currently 140 hydropower projects (large and small) are under construction in the state. Out of these 11 hydropower projects, a total of 2802 MW capacities have been proposed in the Tawang River Basin.[8]

It is important to understand here that the push for development and the need for policy course correction were not only nationally or domestically driven. In the neoliberal India post-1991, the government had envisioned the northeast, as former Minister Jairam Ramesh puts it, the "future of the region lies in political integration with India and economic integration with South East Asia". This transnational vision to connect Northeast with Southeast Asia led the region to undertake a type of mordernisation that was economically driven but strategically contained.(Sarmah, 2016) A region bordered by the countries of Myanmar, Bangladesh, China, Nepal and Bhutan gives it an external dimension, wherein any centre's push (though late) towards building the infrastructure predominantly remained straight-jacketed and determined by India's external relations with its neighbours. (Sarmah, 2016) Thus, the building of hydropower projects in the region should be understood in this pretext. When on one hand China frequently claims to build dams on the Yarlung Tsangpo or Yarlung Zangbo (upper tributary of Brahmaputra) in Tibet,[9] India's attempt to dam Brahmaputra's different tributaries fits in meeting its national energy needs as well as manages the narrative as lower riparian country vis-à-vis China.

Like other anti-dam movements across India, such as the *Narmada Bachao Andolan*, or agitation against the Tehri Dam, and the movement against the construction of the Hirakund Dam, the threats of large-scale displacements and the fear of socio-economic changes also dictate the resistance against hydropower projects in the west Arunachal landscape. But at the same time, the protests in Arunachal Pradesh have expanded vertically to take an international outlook and see the building of dam in corollary to the ebb and flow in the India–China bilateral relation. This expansion is more vivid and evident in the border district of Tawang. Surrounded by Bhutan on west and China on the north, the district of Tawang has been in one of the many arenas of territorial dispute between India and China. This ongoing dispute though takes place at the super-structural level between two countries, its impact is intertwined in the micro-politics of the borderland. Hence even though the internal causes behind the protest in Tawang surround over migration and land rights, the political rhetoric of the protest revolves around China's damming activities across the border.

## The Monpa's Voice of Protest: Causes, Concerns and Fallouts

The two mega-hydropower projects that have been proposed in 2006 and 2007, respectively, are on Nyamjang Chu and other on Tawang Chu rivers (two phases). The Nyamjang Chu Hydroelectric Project[10] with a capacity of 780 MW is designed as a run of the river project, which will harness the hydropower potential of the Nyamjang Chu with a diversion barrage near the Zemithang village, while the Tawang I (600 MW) and II (800 MW) to be built on Tawang Chu River will have a combined capacity of 1400 MW.[11]

The major contestations against the dam constructions have been mainly around the Nyamjang Chu followed by the Tawang Chu. The Tawang Hydroelectric Project is proposed on Tawang Chu River and is located at downstream of fore bridge near Jang (2.5 km downstream of Nuranang Chu powerhouse). The dam site is located about 28 km from Tawang Town on Tawang—Jang road and power house site are located near Yusum village.

It has been obvious that the need for power supply in Tawang has been there. It could be observed that most of the day people are unable to access power for managing their daily chores. That is because Tawang is yet to be connected to the central grid. Electricity in the Tawang villages has come quite late, and the consumption has increased from mainly domestic purposes to commercial purposes with the growing tourism in the district.[12] Even with the increase in demand for electricity, the need for mega-hydel power project with more than 2000 MW capacity was never felt and perceptibly was to be exported for external consumption. Hence when the two mega-projects were proposed, the resistance towards it came from a number of concerns (built around certain cultural and social perceptions) of the Monpas.

The Nyamjang Chu (river) is situated between the valleys of Brokenthang and Zemithang in the Tawang district. This stretch is one of the two wintering sites of the black-necked crane in India and the other being in Sangti Valley in the Kameng district. Apart from serving as the breeding ground for black-necked cranes, the Panchen Valley or the Zemithang is one of the recognised ecological hotspots with the surrounding forests having natural resources like medicinal plant and rare rhododendrons. The people are dependent on these forests for their livelihood and agricultural needs.

There are two concerns about the Nyamjang Chu Dam Project: one is the ecological destruction of the forested areas once the construction starts and merged with the ecological concerns is the religious connotation that would come with the submergence of the breeding ground of black-necked crane.[13] The bird is protected under Schedule I of the Wildlife (Protection) Act, 1972, and is in the vulnerable category as per International Union for Conservation of Nature (IUCN). The National Green Tribunal (NGT) had suspended the environmental clearance granted to Nyamjang Chu Project on the ground that Environment Impact Assessment (EIA) had failed to record the importance of the wintering site of black-necked crane (Grus nigricollis).[14] The 140-km road that would be constructed owing to the project would also affect the 1000 ha of habitat that has been left out in the EIA.[15]

But more than the ecological reason, the religious symbolism of the black-necked crane which is most worshipped by the Monpas is what has bonded the protesters against the Nyamjang Chu dam. As followers of the Gelug Sect of Tibetan Buddhism, the Monpas give a special importance to the black-necked crane as the bird is considered to be a direct reference to the sixth Dalai Lama, Tsangyang Gyatso.[16] A poet at heart Tsangyang Gyatso wrote before 1706, "White crane! Lend me your wings. I shall not fly far". It is after writing similar poems that the Dalai Lama died in house arrest. Ever since, this poem and the crane have remained the spiritual reference to the sixth Dalai Lama.[17] Most of the Monpas in Zemithang believe that the crane to be harbinger of peace and prosperity.

The religious symbolism that started with the Nyamjang Chu protest also goes on to find resonance in the protest against the Tawang Chu hydel power projects. A public perception has been that due to the high location of the dam, the water discharge would submerge the downstream religious and cultural sites of the Monpas, like the water burial site and the shrine of Guru Padmasambhava.[18] The Monpas follow a unique method of disposing dead bodies. One of their burial methods is to cut the body pieces and dispose in the river, wherein the fish would feed on it. This is Monpa's way of giving back to the nature. Similarly, Guru Padmasambhava is one of the worshipped deities of the Monpas, and a submergence of his site would lead to religious discontent among the community.

The resistance though had ecological ground, the religious high ground took precedence when it came to finding a common ground to legitimise their protest. Hence, it was not surprising when a Buddhist monk, Lama Lobsang Gyatso, became the face of the protestors and his organisation Save Mon Region Federation (SMRF) became the legal representation of the petition. In addition for the first time, a monk other than the politician and elite families was seen entering the public and political space[19] in Tawang and debating on the need for the dam as development for the Monpas. The Abbott of the Tawang monastery and the Lamas had until now been spiritual teachers of the Monpas (Gohain, 2017), but this dam protest spearheaded a change in this religious role of the Lamas making it more political.

Another apprehension that runs in the psyche of the Monpas is that of employment and migration.[20] The locals[21] have spoken of outsiders/semi-skilled labourers being brought in for dam construction. These labourers would in turn settle close to the construction site for long period of time. This have raised the concern among the Monpas that such massive labour migration would alter their clan-based settlement pattern and hamper the close knit family. There have been reports of intermarriages among the labourers and the Monpa women which are not always acceptable among the clan.[22] If this continues, many believe it will slowly alter the demography of the Monpas.

Along with the fear for demographical change, the Monpas also fear a loss of land. In the language of construction of indigenousness, the Monpas attach a value system to the natural resources. The Monpas see the forest, rivers and their land as traditional ecological knowledge which is worshipped by them as local deities. The Monpas' extract almost everything from nature to lead their everyday lives ranging from clothing to codification and this they do without hampering the delicate balance that the forest, river system and the land has. For example, they manufacture paper from the pulp of the local *sukshen* tree (which is used in Tawang monastery prints religious books). They manage land and water resources through their traditional institutions in different tier systems. Monpas have distinct categories of forests—clan forest (Tsanpasa, Tsan), privately owned forest (Rangasa), community forest (Mangsa/Borang) and oak-dominated forests having a special place (Ja-sesing). Well-defined rules ensure sustainable use of the forest resources, for example, if a person who is from the same village but from a different clan needs to extract forest resources strictly for local consumption. This upholds the collective and the community use of the forest resources rather than own individual benefits.

Furthermore in the Zemithang area, the whole mountain is considered to be of religious value for the deity Shockseng Gompa, and therefore, extraction of plants from this hill region is avoided. Because of the religious dimensions linked to the species of Rhododendron, Cryptomeria, Juniperus and Cupressus, protection in the name of the local deities is given. There is a fear that land loss due to dam construction would not only alter significantly the Monpas relation with the ecology for subsistence but also led to loss of the forest species which are revered by them as deities. It is important to note here that dependency on ecology is a way of life for the Monpas and protests have become a means to establish this boundary between their socio-religious and ecological habits and the extent of infrastructure modernisation within the community.

The Monpa community have their own way of electing/appointing their leader which has been integrated into the panchayat system, and the leader is now seen as the village headman or the *gaonbura*. Even though the modern political institutions and the social tribal norms coexist among the Monpas, large infrastructure mordernisation at the cost of the ways of life is one of the main reasons propelling the local resistance.

Drawing on the religious and ecological value attached to land, the Monpas are concerned over the types and extent of monetary returns once the land is sold to private companies for dam construction. The EIA report[23] has stated that 80% of the people in the project area were interested in cash compensation and 11% were aware of the displacement.[24] As discussed earlier, the concept of land ownership in the Tawang is twofold: one is private lands, and other is community-owned lands.[25] With increasing need to bring land for big projects like extensive road and energy projects, the private and community lands are acquired. This had led to the emergence of private middlemen or contractors who have become the point of contact with the government and the corporate to distribute the wealth. The middlemen–government nexus has altered the traditional role played by the village headman who was primary mode of contact with the administration for the community.[26]

The *gaonbura* still holds a strong political position, but the emergence of contract class has established multiple chains of power structures (often transactional in nature) which have led the affected people unable to understand whom to approach for fair compensation. The free flow of money has invariably led to corruption while distribution of money. There exists no clear transparency on how the community land will be compensated; hence, the routing of the compensation money becomes easy.[27] In addition, most of the people are agro-pastorals; hence, loss of fertile grazing lands would amount to loss of livelihood and change in employment and settlement pattern of the community. Hence, indirect employments in land contracting, army contracting, tourism and transport business are seen as alternates.[28] This indicates a change in the economic lifestyle of the Monpas.

There is a lack of transparency in the public consultation process before the project was commissioned. As per what was prescribed in the Environmental Clearance Regulations, 2006, public hearing was to be held in New Lumla. Instead, the hearing took place in Lumla of which Zemithang Valley is a part. The distance from Lumla to the affected villages in Zemithang Valley is about one and a half to four hours by car depending on the weather and conditions of the roads. This made it impossible for more than half of the affected households in the region to come to the public hearing. As most do not own a motor vehicle and commute is relatively expensive in the circle, it is only the village headman who attended the hearing representing the affecting people.[29] This in turn did not lead to the direct conveying of the public concern in the hearing. In the public hearing, there is also a huge presence of non-governmental organisations and legal groups alongside *gaonbura*. What is interesting to note is that in the protest against the dam where on one hand there was a religious dimension that pulled the people together, they were also in contact with the NGOs, legal groups and civil society groups for scientific data that would provide the legal sanctity to their protests.

The protests bring out a clear harmony in the flow of information and acceptance among the local Monpas of the empirical and environmental data by the environmentalists and the legalists. At the same time the protesters in the Zemithang valley, especially the SMRF reverted to the religious norms, customs and values attached to the valley thereby making the protest a representation of the collective memory of the Monpas.

This is quite similar to what Social Anthropologist Paul Nadasdy explored in the southwest part of Canada's Yukon Territory where he states that "while to participate in land claims negotiations, they (the local people) had to learn to speak the unfamiliar languages of wildlife management and property law and translate their own understanding of the world into those bureaucratic legal languages". In Tawang, even though a very less number of people understood the legal and ecological jargon, the people still rallied around the ecological cause also due to another factor: Lama Lobsang Gyatso, who many young Monpas consider as knowledgeable and fair, proved to be that one person who brought the environmental concern and the religious concern on one platform, thus bringing all the people together for the anti-dam movement. This concerted working of various groups made for a good legal case, and it became the first hydropower project to be halted on the grounds of environmental protection outside of a protected area.

Over the years, certain mechanisms for compensating local people over the loss of land due to development projects have evolved. Those who lose agricultural lands owned by them as individuals and for which they possess land deeds or *pattas*, are paid monetary compensation or given land elsewhere.[30] However, these over simplistic mechanisms often fail to understand the indigenous relationship that tribals and the community have with vast swathes of land which also houses forests and water bodies. Most community ownership has multiple uses representing multiple heads. The single line of compensatory approach does not always ensure meeting the needs holistically. For example, the compensation packages fall short of the requirements of the upland communities like the agro-pastoralists. In addition, loss of vast lands which might be under farming brings in need to turn the available non-farming lands for agricultural use; in the case of Tawang, it is the forested areas that would be brought under farming so that the community needs are met.

In addition, the Land Acquisition Act of 1894 recognises only individual ownership thereby failing to recognise the community lands among the displaced. One should also note that besides using jhum lands for agriculture, forest resources too are important for medicinal plants and firewood. None of this would be available to the displaced population once displaced. Given the clan and hierarchical-based system of forest use, this displacement will alter the forest usage pattern and access to basic consumption.

This issue is particularly significant while evaluating the social impacts of dams in the Tawang. The submergence of land by hydro-projects will enhance the pressure on the surrounding areas by shortening the jhum cycle, thus affecting the environment and the livelihoods of jhum-dependent communities.[31] Most of the Environment Impact Assessment reports often calculate the submergence of a single type of land, but it does not take into account the change in the landscape, the soil structure and also the land-dependent people.

> A memorandum submitted to the Chief Minister of Arunachal Pradesh by community groups and non-governmental organisations (NGO) working in North Eastern India on 3 September, 2004 after the 'First Consultation Meeting on Planned Dams in Arunachal Pradesh' organised by Arunachal Citizens' Right (ACR) states "the issue of land rights, loss of jhum cultivation areas and loss of customarily owned forest lands are to be seriously examined by the Government of Arunachal Pradesh and the Ministry of Environment and Forests before clearance."[32]

However, such efforts even though taken were not considered while conducting an Environment Impact Assessment.

The above concerns of the Monpas have shaped the protest to a large extent. But it was only after the arrest of the Lama who vigorously campaigned against the dam construction that the protest got its first momentum in the form of a social gathering on 2 May 2016. The subsequent police firing and the death of the two protesters have made Tawang an addition to the incidents of violence and attempts to control the protest by the state machinery in the series of similar anti-dam protests emerging in the rest of northeastern states of India.

## The Differing Dynamics in Dam Protests

### *Emerging Trends in Tawang and Northeast India*

Ever since the development policy in the form of infrastructure building got a renewed thrust in the states of Northeast India, the dam construction on the Brahmaputra basin simultaneously gained momentum. Hydropower projects on the lower Subansiri in Assam, Loktak in Manipur, Dibang and Ranganadi in Arunachal Pradesh and Teesta V in Sikkim started at a full pace which lead to the emergence of various actors with stakes in the project apart from the resource-dependent communities. This led to a conflict of interest on multiple grounds ranging from compensation, connectivity to main national grid, and to displacement, flooding and scarcity in the lower riparian zone. The conflict found its expression in the form of protests breaking out at multiple phases with local and regional support. But more so a trend has emerged that has come to link the anti-dam protests in the northeast of India.

Firstly, to manage all the dam projects there have evolved potential private developers who pay significant amount of money to the governments as advance "before any public consultations, preparation of detailed project reports and receipt of mandatory clearances".[33] This kind of upfront payment puts the government in debt of the private players because now completing the project becomes all the more necessary over any sort of public consultation.[34] The public private collaboration has in turn led to the growth of companies like Mountain Fall India, KSK Electricity Financing, Indiabulls Real Estate and Raajratna Metal Industries who have signed memoranda of agreement with the Arunachal government.[35] In the case of the 1750 MW Lower Demwe Project on the Lohit River, Athena Energy Ventures has committed more than 15% of free power to the state government as royalty, 16% significantly more than the 12% that the public sector NHPC would have given.[36] This has also been the case in Tawang. Prior to the protests in Zemithang, the village headman and the local in the Lumla Circle were aware of the private participation in the building of the dam. Hence, one important point of assertion came when the local Monpas chased the site engineers and the contractors away from the dam site.[37] These tales, continuously reiterated, have found a recognition and acceptance in the collective memory of the protestors.

Secondly, the construction of multiple dams on a river basin has brought in cumulative concerns of migrations, both as displacement and as influx of workforce. For example, one single dam, the 3000 MW Dibang Multipurpose Project proposed on Dibang River in Arunachal Pradesh, requires an estimated workforce of about 5800. But when 17 large dams are proposed simultaneously on Dibang, the work force needed to complete the project doubles to more than a lakh. Thus, the shared concerns of in-migration remain and are not effectively addressed as the workers are often brought by the private contractors with little control by the NHPC.

In addition, when apprehension about submergence of land and habitation grows the apathy towards the outside migrants grows as the indigenous community and then sees them as imposition on what was originally theirs. The 1250 MW Hutong II

Hydroelectric Project proposed on the Lohit River Basin alone in the Anjaw district is estimated to submerge the habitation of half of the population of Anjaw district.[38] As a result, there have been protests against large dams in Lohit River Basin, where affected people had staged road blockade to demand scrapping of controversial dams.[39]

The apprehension about the displacement runs deep among the tribal communities in Arunachal Pradesh due to the presence of the Chakma and Hajong tribes who are resettled in the state as a result of the Kaptai dam built in the Chittagong Hill Tracts of Bangladesh in the 1960s that submerged their traditional agricultural lands. The resettlement of the Chakma and Hajong tribes[40] in Arunachal Pradesh has in turn led to a series of local conflicts between the Chakmas and the indigenous tribal groups. Thus when the Adi tribal group protest against over 2700 MW Lower Siang Project, it is stemming from a similar fear of being uprooted when their wet rice cultivable land would be inundated by the Siang River.

Thirdly, the state not only has come to view the local protests from the lens of security threat but has also provided a counter-narrative that in turn fits the actionable deliverables that helped the state to contain the protest. Hence in the protest against the Lower Subansiri Hydropower Project, an internal memo to the Ministry of Home Affairs warning of a "new 'Red Terror Corridor' along the Assam–Arunachal border" befits this counter-narration by the state. Their labels function as a strategic form of "symbolic characterisation employed against social movements" (Baruah, 2012). Often in this strategic symbolism, external dynamics emerge like when Chief Minister Gogoi spoke of Maoists "infiltrating" the anti-dam protests, he said that "their acts stood to benefit China".

These counteractions serve two purpose one that it brings about a strategic and security ring to the social movements which are often and purely driven by the popular demands for development. Second, the securitisation of the development agenda brings in the need to enforce state machineries in dealing with the resistance movements. Similarly when the protests against the Tawang Chu broke out, the arrests were made based on the use of intelligence. In addition, the central government's construction of Tawang based on its location as the strategic border with China also securitised the protest to an extent that the internal needs of the tribal population become a subtext to the security structure to manage the border communities.

Fourthly, the problems as to why there are resistances to the hydropower projects in the states across northeast are because of the inability by the government to legitimise the necessity of the projects and harmonise the needs of the people with the need for infrastructural development. The legitimacy is two-faced: social and legal legitimacy and political legitimacy. If legitimacy was acute then why in the first place the push for hydropower development took place? It is important to note here that along with the drive to develop, the push for hydropower projects acquired a new lease of life at a time when climate change has moved up on the global policy agenda. The singular need to limit carbon emissions has in turn made hydropower more acceptable as an energy source. Hence, the domestic debate on dam that primarily focused on the environmental fallouts had also been acceptable to the need for renewable source of energy. The "development of rivers and the quest to transform fluvial powers into national assets—hydroelectricity, navigation, irrigation and flood control"—have in turn brought out the multifaceted nature of the dam-building process (Baruah, 2012). The hydropower development is occurring under profoundly different conditions with "state and markets working in collusion, with one exclusion; the acute understanding on the land rights and dependency of the tribal community on the water and forest" (Baruah, 2012).

Lastly, the protests have been moulded and often shaped by historical narratives on whether a dam would be catastrophic or beneficial for the community. The past precedents have played a significant role in creating an indigenous knowledge base with a sense of ownership by the affected community that often rejects or imposes on the scientific reasoning provided by expert committee groups investigating the safety of the dam. Even though legal and technical jargon matters, the perception towards the dam is built on the basis of the history. For example, in Subansiri, eyewitness to the Great Assam Earthquake on 15 August 1950 measuring 8.7 on Richter scale brings back memories of massive landslides and flash floods downstream. The dam constructions may or may not result in a similar disaster, but the past recounting of the disaster without reconciliation sustains the mental vulnerability of community.

For instance, the historical and travel accounts by English botanist Francis Kingdon-Ward documented the earthquake's epicentre in Rima, Tibet. He said some of those massive landslides—"solid-looking hills…in the grip of a force which shook them as a terrier shakes a rat"—blocked the downstream flow of the Dihing, Dibang and Subansiri rivers, and tributaries of the Brahmaputra and when the trapped water burst through in cascades a few days later, it caused catastrophic floods downstream.[41] These are the same rivers on which major hydropower dams are being built or planned today. The earthquake and the floods of 1950 are deeply etched in the memory of the people of the Brahmaputra Valley. After many years, the oral history still remembers the floods as *pahar bhanga pani* (hill-destroying floodwaters) and *bolia pani* (floodwaters driven by madness).[42]

As sociologist Ulrich Beck points out experts can only presume the cultural acceptance of risks, they cannot produce that acceptance (Baruah, 2012). Then what produces that acceptance of risk? The knowledge on whether a dam is risky stems from isolated incidents from the past being quoted. In Tawang, even though a memory of a climatic calamity does not exist it did not discount the protestors from knowing that "history should not repeat itself". The experiences of repeated flash floods and landslides over road constructed bring the fear that more construction could lead to more such events.

The perception of the dam affecting downstream has travelled far in the region to justify the religious undertone to the protest. As Sanjib Baruah puts it, the vernacular knowledge among the local agitators based on first-hand experience has challenged the expert knowledge of the onsite engineers. When talks about clearance to the Nyamjang Chu and Tawang Chu were done, the local body of knowledge on the sacred nature of the breeding ground of the black-necked crane was being corroborated by the expert opinion from NGOs and lawyers. But contentions remained on who is right and who is not about the dam safety. The environmental reasons took precedence in case of stopping Nyamjang Chu Hydropower Project, while ecology in the pretext of religion found resonance in unifying the protest voices against Tawang Chu dam projects.

The pursuit of dam construction by political elites is often part of a strategy to build popular legitimacy through discourses of development and modernity. The decision on whether and where to build a dam is further shaped by the historically sustained power relations between the national and sub-national levels (in this case the central and the state governments) and the institutions used as a tool to maintain the sub-national legitimacy. But it comes at the expense of displacement, disruption of livelihoods or reduced water availability to the community. This makes the local resistances unique as the organisations in the protests are also part of the sub-national political structures trying to create boundaries and accountability in the systemic decision-making. In Tawang, these sub-national structures are though internally driven the debate on the dam constructions has remained externally pivoted.

## The Transnational Character to the Hydropolitics in Tawang

The term "border" is often associated with "conflict" and thereby dealt with a tinge of paranoia and hypersensitivity by the security institutions in India. India's own perception on the borderlands has often determined the type of schemes that is befit for local development without taking into account that infrastructure along the transboundary rivers not only defies the existing boundaries but also poses a challenge to neatly conceptualising interstate relations in general and water-sharing issues in particular.[43] This perception has run in contestation in the northeast of India on what is widely perceived as the centre's lack of resolve in understanding the extent of development and raising the border concerns with China.

It has also not helped that while India and China have signed agreements for the sharing of flood-forecasting data, a lack of water sharing treaty between the two countries has sustained the concerns of both the upper and lower riparian communities as the construction of dams continue. These concerns are further exacerbated by the fact that there are currently no platforms to institutionalise regular interactions between the centre and the northeastern border states on many of the cross-cutting issues like displacement, pollution and floods. The Inter-State Council (ISC), a forum designed to bring all chief ministers to work on operationalising coordination mechanisms between the centre and the states, has only held two meetings so far, the last one being held in December 2006.[44] Centralised agencies such as the Brahmaputra Board under the Ministry of Water Resources have also shown the limits of a top-down model that does not seek to involve active cooperation with the states.[45]

Water itself acquires different social dynamics depending on its usage as resource, commodity, culture and religious to being a geopolitical significance. This different nature of the water means that any attempt to restrict the flow of the river will have impact not only on the community dependent on it for livelihood but also on how the different states (power centres) dispute and cooperate over water. Thus, the politics of dam design and construction on international and transboundary river invariably bring in the politics both positive and negative. As much as there is contention over the rights on the profits of energy distribution, any collateral damage to the environment and downstream pollution are also point of political debate.

In the northeast of India, the symbolic capital of Brahmaputra and its tributaries gets understood not only on the historical precedents of the river's interaction with human within the boundaries but also on how India's neighbouring country understands and uses the river. Noted as the "*downstream dilemma*", India shares Brahmaputra with China in the north and Bangladesh in the south. The interlinkages among the three countries are therefore not only bilateral and territorial but also "*riveral*" in nature. Thus, any activity in the upper riparian zone by China will have an impact if not situational but definitely in the ideational level on how India deals China in its larger bilateral relations.

As the $1.5 billion Zangmu hydroelectric dam, China's largest hydroelectric dam on the Yarlung Tsangpo (upper tributary of Brahmaputra), became fully operational, it brought in a virtual apprehension over its likely downstream impact. While over-interpretation and hysteria have tended to take the place of informed scholarship and media, India's official narrative has largely tended to downplay many of these concerns. The debate has also unwittingly ended up being a single-issue debate, fixated on water diversion.[46]

But in Tawang, the debate has not been on water diversion but of comparison and competition. While at the core the internal issues of socio-economic and religion remain the reason for mobilising the resistance against the dam, the historical memory of the 1962 war in Tawang gives an external interpretation to the protests. These interpretations framed through public writings and media reportages go on to add a dimension that becomes synonymous or one of the concerns of the Monpas themselves that the dam is India's means to balance China in its infrastructural development across the border. Geopolitical attitudes thus disseminate the discussions on international water courses,[47] and an already existing agitation at the sub-state goes on to acquire a geostrategic significance in an interstate dispute.

## Conclusion

India has had examples of multipurpose projects, but in spite of that the government has not been able to build a framework or a statutory body to focus on the development of the Brahmaputra River Basin. What is happening in most of the Brahmaputra Valley and in its tributaries is resistance by the riverine communities refusing to accept the risk definition of power elites bent on pursuing a strategy of accumulation of land with little regard of the impact on the communities.[48]

The local protests in Tawang especially on the two rivers have differed depending on the community's dependence on resources and their relation with the land and the forest. The forest relations and ecology in Nyamjang Chu differ from that of Tawang Chu. Thus when one became admissible in court, the other is yet to be. Second, centre's relation with Arunachal Pradesh in the context of Tawang also determines the progress and intensification of the protest. While protest continues to rage in other parts of Arunachal Pradesh and Assam, verdict in Tawang is one of a kind. With protests spiralling in Arunachal Pradesh, Assam and Sikkim and most importantly in Nyamjang Chu (now stopped) and Tawang Chu Phase I and Phase II, it is a larger message to say that the Monpa's lifestyle would undergo a huge change once the land is used for construction. This fear of change has led to the belief that their culture and social identity are eroding and this is where the rationale of the Monpa's protest against the dams stands.

The need to resolve dam agitation could only be brought out with a common statutory body. The idea of Northeast Water Resources Authority has gone nowhere because of the opposition from the Government of Arunachal Pradesh which "gives no weight to flood control and navigation benefits and gives high weight to any submergence". While the flood control would come as relief to Assam, addressing submergence is a relief to Arunachal Pradesh. But the disagreement between the two states also highlights a deterrent in collective water resource management. The increased pressure on resources, erosion and water diversion plans raise livelihood questions for river-dependent communities both within and across the borders. Given these ripple effects, there is a need for transboundary water governance with both transnational and bilateral frameworks. But as Nimmi Kurian raises, can we frame the debate on water with China in ways that can create institutional entry points for a set of missing issues that are currently invisible to the mainstream policy and research gaze? These protests are a manifestation of these issues that were overlooked for a substantial period.

## Endnotes

1. For more details on the series of events leading to the firing, https://thewire.in/environment/arrested-over-green-tribunals-suspension-of-hydel-project-says-arunachal-anti-dam-activist.
2. The petition to stop the hydropower project was filed by Save Mon Region Federation (SMRF) to the National Green Tribunal.
3. For more, see Lavers and Dye (2019).
4. Baruah (2012).
5. Ibid.
6. Ibid.
7. Bej and Khatoon (Accepted Chapter for Publication).
8. For full report on the number of hydropower projects proposed in the state and the district, refer https://www.livemint.com/Industry/JexN1grDoLQtMgvWYxOx7N/Environment-ministry-panel-defers-clearance-to-600-MW-Tawang.html.
9. For China's claims over Arunachal Pradesh, refer https://www.sentinelassam.com/news/does-history-support-chinas-claim-over-arunachal-pradesh/.
10. Of the 143 projects planned to be built in Arunachal Pradesh since 2001, the Nyamjang Chu Project is the largest of the 13 hydropower projects and was funded by LNJ Bhilwara Group, a textile and steel company.

11. Dutta (2012).
12. The data was gathered during the field visit by the author in Jung, Zemithang and Tawang Circle in April 2018.
13. Bej and Khatoon (Accepted Chapter for Publication).
14. See the judgement by the NGT, 2016, pp. 21–26.
15. Dutta (2012).
16. Bej and Khatoon (Accepted Chapter for Publication).
17. To understand the symbolic importance of the black neck crane, refer https://timesofindia.indiatimes.com/india/hydel-project-in-arunachal-threatening-a-dalailama/articleshow/67170458.cms.
18. Ibid.
19. Lama Lobsang Gyatso decided to contest election against Pema Khandu, his home constituency, but withdrew later. See https://www.ndtv.com/india-news/buddhist-monk-lama-lobsang-gyatso-up-against-arunachal-chief-minister-for-assembly-polls-2011867.
20. Bej and Khatoon (Accepted Chapter for Publication).
21. Based on field interview in April 2018 of the Lumpho village headman, Zemithang, gaonbura and affected families.
22. Based on field interview in April 2018 of the Lumpho village headman, Zemithang, gaonbura and affected families.
23. Apart from the Environment Impact Assessment Report, more details on the cumulative impact of the development plans on Tawang could be gathered from the June 2015 Final Report, "Perspective Plan for Development of Tawang River Basin", submitted to the Government of Arunachal Pradesh by the North-Eastern Hill University, WWF Tezpur, IIT Guwahati and FRLHT.
24. Dutta (2012).
25. Bej and Khatoon (Accepted Chapter for Publication).
26. Ibid.
27. Based on field interview in April 2018 of the Lumpho village headman, Zemithang, gaonbura and affected families.
28. Bej and Khatoon (Accepted Chapter for Publication).
29. Based on field interview in April 2018 of the Lumpho village headman, Zemithang, gaonbura and affected families.
30. Menon (2008).
31. Ibid.
32. Ibid.
33. Vagholikar and Das (2010).
34. Baruah (2012).
35. Deka (2010).
36. Baruah (2012).
37. Based on field interview in April 2018 of the Lumpho village headman, Zemithang, gaonbura and affected families.
38. Baruah (2012).
39. A brief on the protests across Arunachal Pradesh, see https://www.thethirdpole.net/en/2017/01/11/anti-dam-protests-continue-in-arunachal-pradesh/.

40. For more, see https://scroll.in/article/845129/50-years-on-chakma-refugees-from-bangladesh-are-still-denied-citizenship-rights-in-arunachal.
41. Baruah (2012).
42. Ibid.
43. Tripathi (2011).
44. See https://www.livemint.com/Opinion/c9OZI4C1jO0sBdWLDbfGmJ/The-interstate-council-needs-to-be-rejuvenated.html.
45. Kurian (2014).
46. Ibid.
47. Tripathi (2011).
48. Baruah (2012).

## References

Baruah, Sanjib. 2012. Whose River Is It Anyway? Political Economy of Hydropower in Eastern Himalayas. *Economic and Political Weekly* XLVII (2): 41–52.

Bej, Sourina, and Nasima Khatoon. (Accepted Chapter for Publication). Paradox of Development: Emerging Changes and Contestations Among Tawang Monpas in Arunachal Pradesh. In. Dhananjay Tripathi (Ed.). Routledge. (Forthcoming Publication).

Deka, Kaushik. 2010. Dams: The Larger Picture. In *Big Dams and Assam*, ed. Mrinal Talukdar, and Kishor Kumar, 2–9. Guwahati: Nanda Talukdar Foundation.

Dutta, Aparajita. 2012. A *Critique of the Nyamjang Chu Hydro-electric Power Project: Environmental Impact Assessment (EIA) and Environmental Management Plan (EMP)*. India Environtment Portal. http://www.indiaenvironmentportal.org.in/content/365992/a-critique-of-the-nyamjang-Chu-hydro-electric-power-project-environmental-impact-assessment-eia-and-environmental-management-plan-emp/. Accessed August 12, 2018.

Gohain, Swarnajyoti. 2017. Monpas in Tawang: An Anthropological Perspectives (Conference Presentation). In *Focusing the Frontiers: The Borderland, Identity, Perceptions and Imaginings of Monpas of Tawang in India-China Border.* Bengaluru: International Strategic and Security Studies Programme, NIAS.

Kurian, Nimmi. 2014. *India-China Borderlands: Conversations beyond the Centre*. New Delhi: SAGE Publications.

Lavers, T., and B. Dye. 2019. Theorising the Political Economy of Dams: Towards a Research Agenda. *Future DAMS Working Paper 001*. Manchester: The University of Manchester.

Menon, Manju. 2008. Land Alienation Due to Large Hydro-power Projects in Arunachal Pradesh. In *Land, People and Politics: Contest Over Tribal Land in Northeast India*, Sanjay Barbora Walter Fernandes, (Ed.) 128–159. Guwahati: NESRC and IWGIA.

Mishra, Deepak. 2014. Developing the Border. In *Borderland Lives in Northern South Asia*, David N. Gellner (Ed.), 141–158. New Delhi: Orient Black Swan.

Sarmah, Bhupen. 2016. The Cauldron of Conflict: Politics of Peace, Governance and Development in India's North-East. *Social Scientist* 44 (11/12): 15–36.

Tripathi, Narendra Kumar. 2011. Scarcity Dilemma as Security Dilemma: Geopolitics of Water Governance in South Asia. *Economic and Political Weekly.* 46 (7): 67–72.

Vagholikar, Neeraj, and Partha J. Das. 2010. Damming Northeast India. *Briefing Paper,* Kalpvriksh (Pune), Aaranyak (Guwahati) and Action-Aid India (New Delhi).

**Sourina Bej** is a Research Associate in the International Strategic and Security Studies Programme at National Institute of Advanced Studies (NIAS), Bengaluru. Her research interest is on water governance, climate change, identity and development politics in South Asia. She has also worked as a correspondent-cum-copy editor at The Times of India, Chennai.

# Chapter 11
# Arunachal Pradesh in India-China Relations: Trends in Chinese Behaviour and their Implications

**Jabin T. Jacob**

## Background

India's Northeast is virtually landlocked and has only two per cent of its boundaries contiguous with the rest of India with the remaining 98% forming international frontiers with Myanmar, Bangladesh, Nepal, Bhutan, and China. Every state in Northeast India has at least one international neighbour with Arunachal Pradesh having borders with three.

Connected—tenuously, in the minds of Indian policymakers and strategic planners—to the Indian mainland by the Siliguri corridor, it is the security dimension that has historically dominated New Delhi's approach to the Northeastern states. For much of the period following Independence, insurgencies in the region—supported for a significant period of time in the past by China—have limited development options in the Northeast. But, just as significant was New Delhi's almost bipolar approach of simultaneously seeking to preserve the peoples and cultures of the land in some kind of modern-day Eden and continuing the colonial exercise of taming the frontier by force of arms and a legal regime that is markedly different from what obtains in the rest of India.

In the case of Arunachal itself, following 1962, there was the additional blinkered policy of keeping the state deliberately underdeveloped in terms of roads and other physical infrastructure for fear of a repeat of the Chinese intrusion of 1962 or at the very least to slow it down. It has taken the increased pace of development of infrastructure on the Chinese side of the border in recent years and the fact that this is increasingly visible or that news of this has trickled down to the many ethnic groups of Arunachal Pradesh for the Indian central government to begin a process of fast-tracking infrastructure development projects in the state in the mid-2000s.

J. T. Jacob (✉)
Department of International Relations and Governance Studies, Shiv Nadar University, Gautam Buddha Nagar, Uttar Pradesh 201314, India
e-mail: jabinjacob@gmail.com



M. Mayilvaganan et al. (eds.), *Tawang, Monpas and Tibetan Buddhism in Transition*,
https://doi.org/10.1007/978-981-15-4346-3_11

## Flashpoints

There have been several issues of contention between India and China involving Arunachal Pradesh over the years. In this section, three of these are examined briefly—boundary transgressions by Chinese troops, the issue of Chinese visas to Indian citizens from Arunachal, and the Dalai Lama's visits to the state. The next section offers a detailed examination of a fourth and new kind of Chinese sovereignty-building involving Arunachal Pradesh, namely the renaming of places within Arunachal by the Chinese government.

### *Boundary Transgressions in Arunachal*

The issue of Chinese transgressions in Arunachal Pradesh is a long-standing one with reports in the public domain dating back to at least 2003 (*The Times of India* 2003). The simple explanation for Chinese transgressions and the alleged creeping occupation of Indian territory by the Chinese—or 'salami slicing', to use a more current usage—is that as long as there is no agreement between the two sides on the exact alignment of the Line of Actual Control (LAC), neither side can really say that the other is in violation of any agreement or of having intruded into its territory.[1] Transgressions by both sides, therefore, must be seen as a natural fact of life on the LAC between India and China and will continue, whether out of genuine belief that the territory in question belongs to the side that transgresses, out of a need to show presence in areas still infrequently visited by both sides or simply because it is often possible to transgress and get away with it.[2]

However, what is of concern for India is that Chinese settlements and infrastructure development is inching ever closer to the Indian perception of the LAC and that this, in turn, has consequences for local populations in Arunachal Pradesh in terms of a demonstration of good infrastructure on the Tibetan side that is in marked contrast what exists on the Indian side. The other aspect is that such developments will over time include territory that is within India's perception of the LAC, given India's lack of capacity even today to access many areas regularly or because populations have moved inland to the bigger townships in Arunachal.

From a military, and often economic, point of view, it makes a great deal of sense to 'nibble' away at territory as it were, or occupy it steadily, if any side thinks it can get away with it and if it has the capability to sustain its presence, and also because this gives less provocation for conflict than a sudden and large grab of territory would. Given the state of transport and communication infrastructure on the Indian side and the sharply superior progress made in such infrastructure on the Tibetan side, it is but natural that the Chinese will have more cause to continue such activity and appear to have even 'succeeded' as far as Indian perceptions are concerned.[3] China's 'new forward policy'[4] is facilitated by the availability of resources for military modernization and infrastructure development while India, by contrast, continues to be too slow in implementation even if the funds for projects are sanctioned.

## *Chinese Visas to Arunachalis*

Another issue that has often vexed Sino-Indian bilateral ties with respect to Arunachal has been the question of Chinese visas for Arunachali residents. From issuing no visas, China moved to a period of issuing stapled visas—in each case, immigration authorities in India simply did not allow Arunachalis to travel or ended up cancelling official visits or reducing the size of delegations when government officials were involved.[5] While the issue has not hit the media in several years now, there is no reason to think that the policy on either side has changed.

Broadly, it would appear that China has adopted multiple approaches to issuing visas to Arunachalis. Ordinary civilians, including political leaders from Arunachal Pradesh, it would appear can come under three approaches—no visa required, regular stamped visas, or stapled visas. Officials of the central government, however, have either been told they did not need visas or refused a visa altogether.

Meanwhile, one could argue that from saying that no visas were required, the issuing of stapled visas to Arunachalis is possibly, a step forward—an acknowledgement that the area in question is disputed, and by implication, amenable to resolution by negotiations (Jacob 2015).

Meanwhile, New Delhi has sought to undermine any historical Chinese claims to Arunachal Pradesh in a different way through the Republic of China (ROC) government on Taiwan. The ROC officially has never ceased to claim that it is the legitimate government of all China and continues, therefore, to claim all the territories that the People's Republic of China claims and more. This includes territories in dispute with India, too. But successive governments in Taiwan led by the pro-independence Democratic Progressive Party (DPP) have whittled away at these legacies of the ROC vis-à-vis the mainland. A case in point is of how the part of the government dealing with Mongolian and Tibetan issues was first gradually downgraded from a ministry to a commission before finally being dissolved in September 2017 under the present DPP dispensation. It was this, ministry/commission that had dealt with issues related to Arunachal previously. But even before the disbanding, New Delhi and Taipei have cooperated in the case of Indian citizens from Arunachal with the Taiwanese government hosting the latter on its programmes (see, for example, Tapiei Economic and Cultural Centre 2016).

## *Visits of the Dalai Lama to Arunachal Pradesh*

For China's leaders, the domestic implications of protests and instability in its minority areas, particularly Tibet and Xinjiang, are matters of grave national interest—these are called 'core interests'—and foreign involvement in these issues is considered as a sign of the highest hostility. Thus, it is that anything to do with the Dalai Lama, religious head of Tibetan Buddhism and considered the fountainhead of Tibetan resistance to China, brings severe reactions from the Chinese government. Foreign governments hosting him for visits or conferences have often faced Chinese ire and in recent years, given China's growing economic prowess, painful economic sanctions, and political snubs from Beijing.

As far as India is concerned, however, the fact that Dalai Lama lives on Indian soil under government protection makes it particularly odious for sections of the Chinese security and strategic establishment. This is one explanation for the intractability of the boundary dispute, which historically was a boundary between India and Tibet. And yet, it is also true that bilateral relations between India and China have long moved from being held hostage to the boundary issue or the Dalai Lama factor to cooperation on many fronts as well as competition on an ever-widening set of issues.

The Dalai Lama has visited Arunachal Pradesh not infrequently—at least six times before 2017 (Ministry of External Affairs, India. Government of India 2017) and it has drawn Chinese opposition every time. His latest visit in April 2017 was unusual for it was a particularly high-profile and long-drawn-out affair starting with an event in Assam where the Assam Governor Banwarilal Purohit, Assam Chief Minister Sarbananda Sonowal, and several state cabinet ministers shared the stage with him (Chief Minister of Assam 2017) while in Arunachal he was accompanied by the Arunachal Pradesh Chief Minister Pema Khandu (Financial Express 2017). Union Minister of State for Home Affairs, Kiren Rijiju, himself a Member of Parliament from Arunachal Pradesh also accompanied the Dalai Lama for parts of the trip (Singh 2017). The high degree of official involvement not surprisingly drew sharp reactions from Beijing (*Global Times* 2017b).

The Indian government's use of the Dalai Lama to signal to Beijing has been inconsistent to say the least and this applies to both the United Progressive Alliance in power at the centre from 2004-14 and the present ruling National Democratic Alliance. To take the case of the latter, after the big government support to the Dalai Lama visit in 2017, the Foreign Secretary in February 2018 issued a 'classified circular' asking all government leaders and officials to skip functions where the Dalai Lama was going to be present as ties with China were at a sensitive stage (Ghosh 2018).

## China's Renaming Exercise in Arunachal Pradesh

In a notification dated 13 April 2017, China's Ministry of Civil Affairs announced that it was standardizing the names of six places in its 'south Tibet region' (*Zangnan diqu*) that refers to a portion of the Indian state of Arunachal Pradesh. The notification *Guanyu zengbu Zangnan diqu gongkai shiying diming (di yi pi) de gonggao* ('Announcement on the Supplementary Use of Place Names in the Southern Tibet (First Batch)') actually states that it is the 'first batch' (*di yi pi*) of such announcements even though no date for the next round is given nor an explanation forthcoming for the timing of the present one. The announcement is as per the 'relevant regulations' (*youguan guiding*) of the State Council's 'management of place names' (*diming guanli*) (Ministry of Civil Affairs of the People's Republic of China 2017).

The names of the six places in Arunachal Pradesh have now been given 'standardized' names in Chinese characters, and the Tibetan and Roman alphabets with accompanying geographical coordinates. These coordinates have, however, created a degree of confusion in India as at least some early conclusions on what the names referred to have been disproven. While it can be said with some certainty that Wo'gyainling and Mainquka refer to the Urgelling Gompa in Tawang and Mechuka in West Siang district, respectively, the coordinates of Mila Ri, Qoidêngarbo Ri, Bümo La, and Namkapub Ri do not suggest any particular significance. The last two places were originally thought of as Bum La in Tawang and as Namka Chu also in Tawang, that was the site of the first Chinese attack in 1962 but the geographical coordinates do not match (Joshi 2017).

## *Reasons for Chinese Renaming*

One Chinese scholar, Xiong Kunxin, a professor of ethnic studies at the Nationalities University in Beijing, saw the announcement from a legal standpoint as both applying the law of the country and affirming China's territorial sovereignty. Another scholar from the Tibet Academy of Social Sciences, Guo Kefan, argued that the move was part of creating leverage for China in the border negotiations with India (cited in *Global Times* 2017b). The very ordinariness of these reasons—part of what any country would legitimately do as part of the process of extending and developing its administration and capabilities—suggests that this will not be a one-off exercise.

As to the timing, it is an easy inference that the Chinese naming exercise is a response to the visit of the Dalai Lama to Arunachal Pradesh from 7 to 11 April that was for all practical purposes organized by the Indian central government. However, this might not be the most important aspect of the Chinese action. As has pointed out, the Chinese have engaged in such naming exercises previously in the case of India as a way of hardening claims in other sectors of the boundary dispute (Joshi 2017). And by their nature, preparations for such exercises are long in the making and the timing is only a secondary, if not an unimportant, aspect.

There is no doubt that much thought has gone into China's renaming exercise and this is evident in the other processes that it has set in motion in Tibet and with implications for Arunachal.

For instance, there has been greater international access accorded to Tibet in general and which includes areas opposite Arunachal. It might be recalled for instance that China's 4th Forum on the Development of Tibet in August 2014 and the 5th Forum in July 2016 saw foreign participants visiting Linzhi and Shannan prefectures, respectively.[6] China has also taken UN officials to visit Nyingchi (*Tibet News* 2016) while, of course, senior Chinese officials visiting these areas are a common enough phenomenon.[7]

The implication is that the Chinese are beginning to hard sell their position on India–China boundary dispute—hitherto pretty much ignored by the Chinese themselves—to international audiences as well as domestically. This could be both cause and effect of a hardening Chinese position on Arunachal, specifically, and the dispute with India in general.

## *Indian Renaming*

Meanwhile, it is not the case that it is only China that engages in such exercises. Less noticed perhaps and now entirely forgotten in mainland India but certainly not in its Northeast itself are such non-native and Sanskrit-origin names as 'Meghalaya' or 'Arunachal' itself, for instance. In Arunachal, meanwhile, there are several points on the border given North Indian names after soldiers of the Indian army, which have now become part of regular usage. While Hindi forms a link language in Arunachal, it does not mean that multiple ethnic identities (or the attendant rivalries) have been subsumed in an Arunachali, leave alone Indian, identity.

The point is that the Chinese renaming exercise might generate the usual statements in Arunachal about Chinese perfidy but that does not necessarily mean that New Delhi can take Arunachali sentiments for granted. In fact, alongside the criticism of China, there is also an increasing trend to point out the Indian central government's own shortcomings including its inability to settle the boundary dispute quickly.[8]

## Implications of the Chinese Renaming

Among the implications of the Chinese renaming, it might be important to note one of Xiong's statements, namely that the announcement showed that China was gaining a better understanding of the geography in Arunachal Pradesh. This then points to at least two things.

One, there is clearly better Chinese access to geographic and administrative information from inside Arunachal probably the result of better satellite imagery as well as of analysis of open-source information available about developments in Arunachal and locally available intelligence. This in itself is not surprising and is, no doubt, the case for India vis-à-vis Tibet and Xinjiang as well.

However, two, such understanding and analysis need not be limited to only geographic data but also to other aspects of Arunachal including its politics, economy, and society. Interactions with Chinese scholars of India on the sidelines of conferences and academic interactions whether in China or in India do occasionally throw up indications that even as these scholars might not necessarily specialize on the boundary dispute they are expected to be reasonably aware of developments in Arunachal and Ladakh as well as part of their study of India. This is evident in multiple ways.

On occasion, there is a visiting scholar to India who has some experience of having served in areas close to Arunachal Pradesh or Ladakh but who has arrived with entirely different research themes to speak on or study. Another interesting development in recent years has been the willingness of Chinese academic institutions to invite for visiting fellowships scholars from India working in either the border regions or on issues related to the borders.

By contrast, Indian scholars of China or even retired government officials engaged in China analysis usually have little or no understanding of even terrain or geographical conditions in Arunachal Pradesh and Ladakh, leave alone local politics and social issues in these areas. While there are anthropologists, sociologists, and other social scientists who do study Arunachal Pradesh and Ladakh deeply, their input into the analytical and policymaking process at the national level with respect to their field areas themselves and to the India–China relationship appears limited.

There certainly is little to no coming together of these studies with those of Tibet- and Xinjiang-related works within the government in any meaningful manner. The Indian army and the intelligence services probably are sites where some of this coming together happens but structurally these are institutions that also lack the wherewithal and influence required for any consistent or long-term work based on deep, scholarly research and inputs and, are therefore, driven by more immediate functional needs. It is a reasonable surmise, meanwhile, given the record, that there is much better integration of the analytical and policymaking arms in the Chinese system. While there are issues with efficacy in the Chinese system, too, the fact that such structures are in place can only mean forward movement and improving efficiency as opposed to having no structures at all in place that could create the necessary synergy.

Meanwhile, there is a need for greater interactions within the academic disciplines in India itself that are increasingly finding reasons to work on China but are not yet able to contribute coherently to shared knowledge production or to government policymaking, including in cooperation with scholars of Arunachal and other Indian states bordering Tibet. As to the latter aspect, it is the government's own tight-fistedness and inefficiencies that ensure that Indian scholars lack the resources required to carry out their research and therefore, remain largely out of the loop of inputs going into government analysis and policymaking.

## Conclusion

It must be remembered that while for the Chinese, Arunachal Pradesh may today be at the heart of the Sino-Indian boundary dispute, the dispute itself is increasingly less at the heart of the Sino-Indian relationship. And, this appears increasingly the case also for India, even as it does not countenance any territorial concessions or disturbing of settled populations, and remains in general, far more sensitive to the persistence of the boundary dispute than China is.

It is worth noting that Indian Prime Minister Narendra Modi's visit to Arunachal Pradesh in February 2019 did not draw as sharp a reaction from the Chinese side as usual. While China's Ministry of Foreign Affairs spokesperson did reiterate that 'The Chinese government has never recognized the so-called "Arunachal Pradesh" and is firmly opposed to the Indian leader's visit', she also called on India to 'cherish the momentum of improving relations between the two countries' (Ministry of Foreign Affairs, the People's Republic of China 2019), suggesting a new sensitivity to the state of the relationship. Of course, this might be interpreted in terms of the so-called 'Wuhan spirit' arising out of the supposed understanding achieved by both sides at the 'informal summit' between Modi and Xi in the central Chinese city in April 2018. The summit was necessitated by tensions in the relationship following the more than 70-day long stand-off in mid-2017 between the two countries on a tiny piece of Bhutanese territory that the Chinese had tried to illegally build road infrastructure on. But the more likely reason for the rather gentle tone by Chinese standards was probably because the Chinese were still under pressure in the wake of an ongoing trade war with the Americans.

Overall, the Sino-Indian relationship has been in something of a downward spiral for several years now—periods of apparent improvement appear largely temporary exercises to ensure that state or other visits go off without a hitch and contribute at least good optics. In practical terms, there seems to be little trust between the two sides. The Chinese are focused on promoting their strategic Belt and Road Initiative of infrastructure development and Xi Jinping's own prestige on global forums while lack of domestic capacity has made it more attractive for New Delhi to partner closely with the USA and its allies in a more or less open challenge to Beijing' interests. Given this state of affairs at the bilateral level and internationally, whether the issue of Arunachal Pradesh will now re-emerge, be refashioned or disappear altogether from the radar of India–China bilateral relations, remains to be seen.

## Endnotes

1. Former Indian Army chief, Gen. Deepak Kapoor, stated as much in an interview on television (*The Indian Express* 2008).
2. At least on the Indian side, there are several vast stretches along the LAC in Arunachal Pradesh that see Indian troop presence only once a year during the long-range patrols undertaken by the Indian Army and the Indo-Tibetan Border Police (Author's interviews in 2011).
3. Author's interviews in 2008 and 2011.
4. This expression for China's infrastructure development in and increasing access to border areas was first used in Jacob (2011: 135–139).
5. For more on this aspect, see Jacob (2015).
6. For more on this aspect see Bhutia (2016).
7. For Shannan Prefecture, see for example, China Central Television (2015).

8. Author's interviews in 2008, 2011, and 2013.

## References


Bhutia, Tshering Chonzom. 2016. A Review of the 2016 Forum on the Development of Tibet, *ICS Analysis*, No. 42, August. http://www.icsin.org/uploads/2016/10/10/2ecbc04c6d6a4531bebb84cf4c7dafe7.pdf.

Chief Minister of Assam. 2017. Dalai Lama at Namami Brahmaputra Festival: Understanding the Value of Religious Teachings Important for Peace, *Press Release*, 2 April, http://cm.assam.gov.in/pressrelease.php?displayid=100375.

China Central Television. 2015. Senior Chinese Official Visits Shannan Prefecture of Tibet, 11 September. http://english.cntv.cn/2015/09/11/PHOA1441954970127688.shtml.

*Financial Express*. 2017. Dalai Lama Holds Preaching Session in Arunachal Pradesh's Bomdila; CM Pema Khandu Attending the Event, 5 April. http://www.financialexpress.com/india-news/dalai-lama-holds-preaching-session-in-arunachal-pradeshs-bomdila-cm-pema-khandu-attending-the-event/615482/.

Ghosh, Abantika. 2018. Govt Sends Out Note: Very Sensitive Time for Ties with China, So Skip Dalai Lama events, *The Indian Express*, 2 March. http://indianexpress.com/article/india/govt-sends-out-note-very-sensitive-time-for-ties-with-china-so-skip-dalai-lama-events-5083430/.

*Global Times*. 2017a. India's Use of Dalai Lama Card Tactless, 6 April. http://www.globaltimes.cn/content/1041035.shtml.

*Global Times*. 2017b. Standardization of Terms Aimed at Reaffirming Sovereignty: Experts, 18 April. http://www.globaltimes.cn/content/1042979.shtml.

Jacob, Jabin T. 2011. For a New Kind of 'Forward Policy': Tibet and Sino-Indian Relations. *China Report* 47 (2): 135–139.

Jacob, Jabin T. 2015. Arunachal Pradesh in the Sino-Indian Boundary Dispute: Constant Claims, Changing Politics. In *Voices from the Border: Response to Chinese Claim over Arunachal Pradesh*, ed. Gurudas Das, C. Joshua Thomas, and Nani Bath, 48–62. New Delhi: Pentagon Press.

Joshi, Manoj. 2017. Why Is China Renaming Seemingly Unimportant Places in Arunachal Pradesh? *The Wire*, 24 April 2017. https://thewire.in/127790/india-china-arunachal-pradesh/.

Ministry of Civil Affairs of the People's Republic of China. 2017. Guanyu zengbu Zangnan diqu gongkai shiyong diming (di yi pi) de gonggao' [Announcement on the Supplementary Use of Place Names in the Southern Tibet (First Batch)], No. 404, 13 April 2017. http://images3.mca.gov.cn/www/file/201704/1492141901703.pdf.

Ministry of External Affairs, India. Government of India. 2017. Visit of His Holiness the Dalai Lama (HHDL) to Arunachal Pradesh, *Press Release*, 4 April 2017. http://mea.gov.in/press-releases.htm?dtl/28301/Visit_of_His_Holiness_the_Dalai_Lama_HHDL_to_Arunachal_Pradesh.

Ministry of Foreign Affairs, the People's Republic of China. 2019. Foreign Ministry Spokesperson Hua Chunying's Remarks on Indian Prime Minister Modi's visit to the East Section of the China-India Boundary, *Press and Media Service*, 9 February 2019. https://www.fmprc.gov.cn/mfa_eng/xwfw_665399/s2510_665401/2535_665405/t1636521.shtml.

Singh, Vijaita. 2017. Dalai Lama's Arunachal Visit Purely Religious, Says Rijiju, *The Hindu*, 2017. http://www.thehindu.com/todays-paper/dalai-lamas-visit-is-religious-says-rijiju/article17820121.ece.

Tapiei Economic and Cultural Center. 2016. Delegation of Arunachal's Student-Scientist Calls on TECC in India, *Events*, 28 January 2016. https://www.roc-taiwan.org/in_en/post/1759.html.

*The Indian Express*. 2008. Chinese Incursions Due to Different Perception, 23 February 2008. http://www.expressindia.com/latest-news/Chineseincursions-due-to-different-perception-India/276320/.

*The Times of India*. 2003. Chinese Transgressed on LAC in Arunachal: India, 24 July 2003. http://articles.timesofindia.indiatimes.com/2003-07-24/india/27211569_1_lac-chinese-side-chinese-officer.

Tibet News. 2016. UN Representatives Visit Nyingchi in Tibet, 1 June 2016. http://eng.tibet.cn/news/1464741395695.shtml.

**Jabin T. Jacob** is Associate Professor at the Department of International Relations and Governance Studies, Shiv Nadar University, India and Adjunct Research Fellow at the National Maritime Foundation, New Delhi. He was formerly Fellow and Assistant Director at the Institute of Chinese Studies, New Delhi and Associate Editor of the journal*China Report*. Jacob holds a PhD in Chinese Studies and has spent time as a student/researcher/faculty in Taiwan, France and Singapore. His research interests include Chinese foreign policy, Sino-Indian border areas, Indian and Chinese worldviews, and centre-province relations in China.

# Chapter 12
# Border Community and Border Management: A Case of Tawang Tract

**Pushpita Das**

## Introduction

The Tawang tract, owing to its geographical location, is of immense strategic importance to India. Wedged between Bhutan and the rest of Arunachal Pradesh, the tract contains several easily traversable trade routes running in a north–south direction, which provides easy access to the Assam plains and the Indian hinterland beyond.[1] In the case of hostility with China, as was observed during the 1962 border war, occupation of this tract provided China with an opportunity to exert its influence on the local populace as well as Bhutan. Further, the trade routes were exploited by the Chinese Army to enter into the Indian territory, bypass the Indian defences, and cut-off their logistics and communication lines from the rear.[2] Even China's claim on Arunachal Pradesh, which it refers to as "South Tibet", is based on Tawang arguing that Tawang had always been under the cultural influence and administrative jurisdiction of Tibet. It also cites the fact that the 6th Dalai Lama, Tsangyang Gyatso, was born in Tawang.

According to China, since Tibet is part of China and since Tawang was part of Tibet, by extending that logic, Tawang and the whole of Arunachal Pradesh belong to China. The importance of Tawang in the Chinese scheme of claims can be gauged by the fact that China had expressed its willingness to make concessions in the eastern sector if the Indian government concedes its claim on Tawang.[3] India had naturally rejected the suggestion saying that Tawang is an unalienable part of India.

Given the strategic importance of the Tawang tract to India and given China's assertive claim on it "as the minimum condition to consider a settlement",[4] it is imperative for India to secure Tawang and its population to safeguard its own national security and territorial integrity. In this respect, India has been bolstering its defences in the Tawang tract by deploying troops, building strategic road networks for easy

P. Das (✉)
Institute of Defence Studies and Analysis, New Delhi 110010, India
e-mail: pushpitad@yahoo.com



M. Mayilvaganan et al. (eds.), *Tawang, Monpas and Tibetan Buddhism in Transition*,
https://doi.org/10.1007/978-981-15-4346-3_12

mobility of military and equipment as well as implementing various programmes to develop social and economic infrastructure for the people residing in this remote border area. Defence and development together with the active cooperation of the border residents is the strategy employed by India to effectively manage this section of its international border with China.

This chapter begins by discussing the nature of the India–China border and the border space of Tawang. It then investigates reported attempts by external forces to alienate the people of Tawang. In the third part, the chapter analyses the connection of the people to border security and management, and finally, it concludes with suggestions to enhancing community participation in border management.

## The Nature of the India–China Border

The India–China boundary in Arunachal Pradesh stretches for 1126 km[5] from the Bhutan–India–China tri-junction in the west to the India–China–Myanmar tri-junction in the east. The border is disputed by China as it does not recognise the McMahon Line as the international boundary and claims the whole of Arunachal Pradesh, especially Tawang, as its own. With the Se La mountain range separating Tawang tract from the rest of Arunachal Pradesh, Tawang has always been oriented geographically, politically, and culturally to Tibet since historical times. It was the desire of the British-India government to have a secure frontier in the northeast that led to the legal incorporation of Tawang into India in 1914.

British–India wanted to have a frontier that followed the watershed principle, but it realised that Tawang was on the Tibetan side of the watershed. Tawang's close proximity to the Assam plains was considered to be too dangerous for it to be left as a part of Tibet. So, British–India decided to incorporate Tawang as part of the Indian Territory and negotiated with the Tibetans a frontier that would follow the watershed north of Tawang. Tibet agreed to the proposal and ceded the Tawang tract to British–India during the Shimla Conference through an exchange of notes from 24 to 25 March 1914.[6] Thus, the Conference which was convened to define Tibet's frontiers also delineated the international border between India and Tibet, popularly known as the McMahon Line. Unfortunately, the British could not decisively enforce the new boundary on the ground as they became preoccupied with the First World War. As the war broke out immediately after the Conference, other subsequent political developments in India sought their immediate attention.

There were, however, some brief attempts by the British-India government in the later decades to physically register its presence in this region. The first effort was made in April 1938. A Chinese map, published in 1936, showed the whole of Arunachal Pradesh as Chinese territory. In response to the map, Captain G A Lightfoot, the Political Officer in Charge of the Balipara Frontier Tract, undertook a mission to Tawang and informed the Tibetan officials that Tawang was Indian Territory and not Tibetan. Upon his return from the mission, Lightfoot recommended the permanent stationing of British officials in Tawang to assert British sovereignty. This suggestion of Captain Lightfoot was unfortunately ignored by the British-India government.[7] A second attempt was made in 1944 when some administrative and army outposts were set up in Dirang Dzong, south of Se La.[8] In sum, while on paper Tawang belonged to India, on the ground nothing changed as the Tibetan government continued to exercise their religious and administrative control over Tawang by collecting taxes and nominating the head lama of the Tawang Monastery.

Even after gaining independence, India did not enforce the boundary on the ground and for close to four years, Tawang continued to be under the Tibetan administration. It was only after the occupation of Tibet by the Peoples' Liberation Army (PLA) in 1950 that the Indian governments decided to occupy Tawang. So in January 1951, Major Bob Khating, the Assistant Political Officer (APO), set out to claim Tawang as Indian Territory. On the "third day of the Iron-Hare Year, corresponding to 9 February 1951", Major Khating formally announced that the administration of Tawang and its Monastery would be, hitherto, the responsibility of the Indian government. He further stated that Tsona Dzongpon and other Tibetan officials could no longer have any authority over villages south of Bum La and should cease collection of taxes and tributes.[9] China did not protest to the establishment of Indian administration over Tawang at that time.

The flight of Dalai Lama from Lhasa to India through the Tawang tract in March 1959 changed the regional geopolitical scenario as well as India–China relations forever.[10] Tensions between India and China increased manifolds, and China formally claimed Arunachal Pradesh in a letter written by Zhou Enlai to Jawaharlal Nehru dated 8 September 1959.[11] Chinese PLA also started transgressing into the Indian Territory in order to assert China's claim line on the ground. These claims and transgressions across the international border further worsened bilateral relations and culminated in the border war of 1962. The boundary between India and China that came into existence after the war is a military held line and known as the Line of Actual Control (LAC).

Since 1962, China has consistently maintained its claim over Arunachal Pradesh and steps it up occasionally to put pressure on India. For instance, in 1985 during the fifth round of official level talks between the two countries, Chinese vice Foreign Minister Liu Shuqing went back on the "swap deal" and reiterated Chinese claim in the eastern sector. Again, in November 2006, the Chinese Ambassador Sun Yuxi said, "In our position, the whole of the state of Arunachal Pradesh is Chinese territory. And Tawang is only one of the places in it. We are claiming all of that. That is our position".[12] Thus, the border dispute between India and China remained intractable.

The disputed nature of the boundary makes it extremely difficult to manage. As the boundary line is not defined, it results in difference of perceptions between India and China regarding the actual position of the boundary on the ground. These differences in perceptions about the boundary make patrolling of the border difficult and invariably result in transgressions into each other's territories creating further tensions.[13] For instance, there were 273 Chinese transgressions in 2016 which increased to 426 in 2017.[14] In fact, PLA's transgressions in the Sumdurong Chu area in 1986–87, Depsang in 2013, and Chumar in 2014 resulted in the escalation of tensions between the two countries leading to stand-offs between the two armies for months.

Even though both the countries tried to resolve the occasional flare-ups at the border through established mechanisms such as border personnel meetings, flag meetings, meetings of Working Mechanism for Consultation and Coordination (WMCC), as well as through diplomatic channels, the fact remains that the disputed nature of the border is a major source of tension between India and China.

## The Nature of Border Space of Tawang

India's border areas are plagued by pervasive underdevelopment resulting from difficult terrain, lack of transportation and communication connectivity, infrastructural development, and small and unproductive economic base. As a result, people living in border areas do not have access to basic essentials such as food, potable waters, medicines, markets, and employment opportunities. The situation is so grave that it has forced many residing in the high and remote regions of the border areas to migrate to other commercial and urban centres situated at the foothills resulting in depopulation of the border areas. Such depopulation does not augur well for the country, especially in the case of India–China border area which is disputed, because the physical presence of population is necessary to establish India's claim on the disputed territory. The case of Tawang is no different.

Tawang is a district of Arunachal Pradesh surrounded by Tibet in the north, Bhutan in the west and south, and the West Kameng district of Arunachal Pradesh in the east. The district is spread over 2000 $km^2$ and its terrain is characterised by lofty ranges and towering peaks covered with snow, craggy hills with subtropical and alpine forests, steep rugged valleys, and cascading rivers. The average elevation of the region ranges between 3500 and 22,000 feet, and habitation is found in the lower elevations with a cool temperate climate.[15]

Administratively, Tawang is divided into three subdivisions, namely Tawang, Jang, and Lumla. Tawang has a population of 49,997, of which 77.6% resides in rural areas and 22.4% in the urban areas. There is only one urban centre, i.e. Tawang town. The district has a population density of 23 persons per $km^2$ and sex ratio of 714.[16] Although total literacy rate of the district is 59%, the level of education is very low. It ranges between below primary level to higher secondary, and graduates comprise less than one percent of the total population. There is one district hospital with 54 beds, but it has no surgeons or anaesthetists. Access to even this small hospital is difficult because of hilly terrain and scattered habitations. Although electricity connection is there in all the villages, its availability for consumption is for less than half a day. Basic necessities like regular markets, general shops, fair price shops, etc., are also not located at an easily accessible distance for majority of the villages.[17]

As far as connectivity is concerned, the district is connected to Tezpur in Assam by an all-weather road, and all the circle headquarters are also connected by road. In fact one of the earliest roads in Arunachal Pradesh was the Bhalukpong–Tenga–Bomdila–Tawang Road built by the Border Road Organisation (BRO) in 1960.[18] Although the transport and communication system has been developed considerably in Tawang during last decades, but due to frequent landslide, soil erosion and heavy snowfall, and poor engineering works, roads get damaged. Poor road communication is the hallmark of this border area which impedes easily mobility of men and material, especially to the China border.[19] Besides, the nearest railway station to the district is at Bhalukpong, and the nearest airport is Tezpur. There are 3 (three) helipads located at Tawang, Lumla, and Zemithang for landing of helicopters, but most of time they are at the mercy of the mountain climate making landing and take-off period extremely limited.

Tawang's economy is based on subsistence agriculture and animal husbandry, i.e. cultivation of food crops, commercial crops, and rearing of animal like yak, sheep, goat, etc. Handloom and handicraft such as carpet weaving, wood carving, handmade paper making, thangka painting, agarbatti making, and pottery as well as tourism industries also contribute to the economy of the district.[20] While more than 70% of the population is engaged in farming, the majority of people are also engaged in miscellaneous activity such as road construction and driving. Lack of opportunities has forced people to migrate in search of jobs to outside their village leading to casualisation of labour.[21] It is important to note that India–China border is a hard border. No cross-border interaction whatsoever—be it trade or movement of people—is officially allowed along the China border in Arunachal Pradesh.

Given the relative underdevelopment, poor social and economic infrastructure, and poor quality of life in Tawang, it is but natural to conclude that residents of Tawang should nurse a sense of neglect which can gradually transform into a sense of alienation from the Indian state. And if such feelings are simmering among the people of Tawang, it could jeopardise the security of the country.

## Signs of Alienation and Demand for Autonomous Council

Although there are no visible signs of alienation among the people of the Tawang, it is important to remember that the border tract is underdeveloped with the absence of proper infrastructure, industries, or avenues of employment. It is also deficit in housing, healthcare facilities, electricity, and access to other basic amenities. This pervasive underdevelopment has, indeed, created a sense of neglect among the people of the district. This sense of neglect is aggravated when they see well-developed infrastructure just across the border in the form of well-developed all-weather roads, better telecommunication facilities, pucca residential buildings, easy access to health, education, and other essential amenities provided by the Chinese government to its border people.[22]

In fact, frustrated at not being paid enough attention by the state administration, the Monpas has started demanding the creation of Autonomous District Council (ADC) for the Mon region comprising the districts of Tawang and West Kameng under the Sixth Schedule of the Indian Constitution to preserve and promote their "unique and diverse culture and religion".[23] The memorandum for the demand for a Mon ADC reads

> As Monpas, Buddhist by religion; have distinct age-old culture, script, culture and tradition, the demand aimed at preserving their culture, tradition, script and language through self-governance as it would help frame need-based plans and policies and ensure its proper implementation while facilitating blending of traditional system of governance appropriately with modern methods within the provisions of the Constitution. It should not be construed as an act of secession from the state but as an effort to develop a mechanism for proper development of the Monpas through self-governance.[24]

The Mon Autonomous Region Demand Committee (MARDC) argues that the two districts are relatively backward because all development-related decisions are centralised in the faraway state capital. As a result, the far-flung areas such as Tawang and West Kameng do not get adequate attention from the state administration. It emphasises that all-round development requires decentralised decision making and participation of the people in planning, execution, and monitoring of all projects, which could be provided only if an autonomous district council is created in the region. Though the state assembly passed the resolution for the creation of a Mon ADC, the union government is yet to give its go ahead.[25] Here, it is important to note that Arunachal Pradesh is not covered under the Sixth Schedule of the Indian Constitution.

Incidentally, the structure of decentralised decision has existed in Arunachal Pradesh since long. In fact, the state was one of the first where Panchayati Raj Institutions (PRIs) for the decentralised decision were implemented in December 1968.[26] Given that grass-root governance is available in the form of PRIs, the demand for an autonomous district council by Monpas defies logic. More so, when funds from the union government are transferred directly to the gram panchayats unlike the ADCs where funds are routed through the state governments. So, is the demand for autonomous district council more a reflection of Mon identity assertion and less of the issue of lack of development? The answer lies in the fact that examples of assertion of Mon identity can be seen in Tawang and West Kameng. In these two districts, places are being renamed according to their original names and written in the Bhoti script.[27] The Monpas had also raised the demand for inclusion of the Bhoti language in the Eight Schedule of the Indian Constitution.[28]

The drumming up of the Monpa identity, however, should be viewed more as an outcome of local politics rather than any sinister ploy of secessionism. The presence of powerful Monpa personalities such as the late Dorji Khandu and his son Pema Khandu and the late Tsona Gontse Rinpoche helped the Monpa cause as they were successful in broadcasting their demands more forcefully and effectively. In fact, it was under Dorji Khandu that the demand for a Mon ADC gained more traction, and it was during his chief ministership when the state government created the Department of Karmik and Adhyatmik (Chos-Rigs) Affairs (DoKAA) in August 2009 and provided it with special developmental packages to give special attention to the districts of Tawang and West Kameng.[29]

Whatever may be the case, the government needs to be aware and sensitive to these demands because this professed sense of neglect might transform into alienation if concerted efforts are not taken to bring about the overall development of the border areas. At the same time, it has to be careful towards promoting too much of identity politics in the strategically sensitive region lest it backfires. The aim should be to foster in the local people a greater sense of belonging with India. The urgency is felt even more now given that the Chinese President Xi Jinping has turned his government's focus on the development of the border areas of his country. In a recent dispatch, Xi Jinping had told herdsmen in Tibet to settle near the border and stay put to safeguard Chinese territory.[30] He hoped that the settlers will motivate more herders to set down roots in the border area and become guardians of the Chinese territory and constructors of a happy hometown.[31] In other words, the Chinese government has turned its attention towards border residents and it is now eliciting their loyalties through the strategy of development of the borders.

In contrast, India has always discouraged settlements near the border in the name of security and sensitivity. As a result, in vast swathes of the border in Arunachal Pradesh there are no settlements. This in turn has detrimental consequences on the security of the border. The Indo Tibetan Border Police (ITBP) personnel who guard the border have to walk for weeks with ration and other paraphernalia to conduct one long range patrolling as there are no villages near the border to provide them with food and shelter.

## People of Tawang and Their Role in Border Management

Border management has the twin objectives of improving security and facilitating trade and transit, at the same time, preventing the entry of goods and people who are considered to be detrimental to the domestic territory and population. In the case of Tawang, border management involves the security, defence, and development of the border. The aspect of facilitation of travel does not arise because no cross-border interaction—be it trade or travel—is officially allowed across the India–China border in this sector. In order to better manage the border areas in Arunachal Pradesh, the government has deployed the Indian army as well as the ITBP to defend and secure the border. It has also implemented a series of road building projects, besides launching various developmental schemes for the border areas. The government has been actively seeking the cooperation of the border people.

The role of border community in safeguarding the border areas is immense. Sustained and improved relations between security forces and the local community result in better border management practice based on intelligence-based interdictions model.[32] The model states that since the local communities have intimate knowledge about the terrain, movement of suspicious people and their meeting places, as well as other security-related issues, they are often co-opted to provide relevant information/intelligence and act as the eyes and ears of the security agencies deployed in the border areas.[33] They are also the biggest bulwark against any aggression by the enemy as they live there and watch every movement of the military personnel stationed across the border. That is why world over governments encourage people to settle along the border.

According to the 2011 census, Tawang has a population of approximately 50 thousand with 70% being scheduled tribe. Monpas belonging to the Bodic group of the Tibeto-Burman family are the dominant tribe of Tawang. They are believed to have migrated from Bhutan some two centuries ago. There are 235 total villages in the district, out of which 218 are inhabited by the Monpas, who belong to the Lamaistic School of Tibetan Buddhism of the Mahayana sect.[34] The Tawang Monastery is one of the most important elements in the social and religious life of the Monpas. The Monpas are religiously and cultural closely associated with Tibet, but socially they do not consider themselves as Tibetans. This argument is evidenced by the fact that the Monpas do not mix with the descendants of the Khampas who were brought into Tawang to construct the Monastery during the seventeenth century. The Khampas reside in separate villages and are not allowed to own land.[35]

After independence when the Indian government sent Major Khating in a mission to establish an administration in Tawang, the local people welcomed him wholeheartedly. In fact, they cooperated enthusiastically with Major Khating's mission by supplying logistics and information about the area. One of the major reasons for such a welcome was that the Tibetan administration in Tawang was harsh. According to local accounts, the Tibetan taxed locals according to their land holdings, made them work for free, and took away their agricultural produce.[36] If some people failed to pay the taxes, they would be jailed or beaten with canes. There was also a strong suspicion among the Monpas that the Tibetans looked down upon them. Understanding these sensitivities of the local people, when Major Khating established administration, he announced that the Indian administration would not interfere with religious practices of the Monpas, will not practice forced labour, and pay for everything. Major Khating's administration also helped the local victims of the earthquake of 1950 by distributing rations, clothes, and medicines and organised parties and treated the village elders respectfully to win them over.[37]

The people of Tawang welcomed the Indian administration and remained steadfastly loyal to India even through the Chinese occupation in 1962. The credit for the display of loyalty by the local people, who had experienced Indian administration only for a decade, goes entirely to the policy backed by the then Prime Minister Jawaharlal Nehru, his tribal affairs advisor, Verrier Elwin, and the officers of the Indian Frontier Administrative Service. The policy was to allow the local people to retain their customs and tradition and develop according to their genius, respect Tribal rights on land and forests, build up a team of local people for administration and development, and dissuade outside influences.[38] Consequently, missionaries were not allowed to preach and proselytise, and movement of people from the rest of India into the North Eastern Fortier Agency (NEFA) was restricted through the retention of the inner line permit system. The non-interference policy was greatly appreciated by the tribal people, and they looked upon the Indian administration as a benevolent one.

In contrast, the Monpas were extremely wary of China because they had heard horror stories about Chinese atrocities on Tibetan people—how their monasteries were destroyed and monks were either, killed, incarcerated or pauperised by the Chinese PLA.[39] Therefore, in spite of Chinese anti-India propaganda during the

1962 occupation of Tawang, the Monpas did not rise in revolt when the Indian administration was restored even though the Chinese had told them that they would return and that the locals should help them when they came.[40] Also a number of Tibetan porters who had come with the PLA had apparently warned the locals against believing in the Chinese good behaviour. The final straw on the proverbial back of the camel was the flight of Dalai Lama from Lhasa into India. The passage of His Holiness through their territory had a profound impact on the hearts and minds of the Monpas, and they decisively turned against the Chinese.[41]

While the local people remained loyal to India and did not support the Chinese during the 1962 war, there was a perception among the Indian intelligence and security establishment that the Monpas did not go out of their way to help the Indians. Rumours were rife that the local people had helped the PLA by providing them with intelligence and logistical support. As a matter of fact, the Chinese were indeed more successful in developing an intelligence network in the Monpa areas through their agents who entered India in the guise of refuges as well as closely questioning the Monpa traders who went across the border about troop deployments, their stations and outposts, etc.[42] To prevent such future occurrences, the Government of India, as a short-term measure, set up a number of fair price shops along major routes leading to Tibet so that the basic requirements of the people could be met locally which would prevent traders from crossing the border and interacting with Chinese agents.[43]

Taking note of the situation, Nehru argued that it is of urgent necessity to integrate the border areas with India to safeguard the country against future Chinese attacks or its efforts of subversion. He advocated that "the entire frontier people should be ideologically and physically built up to a high state of preparation so that they could confidently face any future Chinese onslaught; and even if the Indian army was compelled to withdraw from these areas, the local population would no longer be passive but carry on active opposition against the Chinese in every village".[44] To give shape to this policy, the Special Service Bureau (SSB) was established in 1963 with the objective to generate a sense of national belonging, security, and vigilance among the border population. For the purposes of training villagers to defend their own villages and if the situation demanded, to participate in a "stay behind role" for the nation, various village level training programmes and refresher training courses in civil defence were started.[45] Besides their core duties of village defence and resistance, the trained volunteers also undertook civic welfare programmes such as teaching children in schools, conducting medical camps, and organising vocational training courses and became the eyes and ears of the SSB.[46]

Similar programmes under which Home Guards units organised as part of the Lok Sahayak Sena also functioned under the NEFA administration.[47] Villagers were trained in the use of small arms, the art of self-defence as well as patrolling tactics to develop a spirit of resistance. The idea of recruiting local tribesmen as future defence volunteers highlighted the immense confidence that the Indian government had on the loyalty of these people. These programmes were effective in considerably reducing Chinese intelligence agents prodding along the border. Furthermore, the success of the programme can be gauged by the fact that during the 1965 war with Pakistan when China threatened to attack, around 800 volunteers were mobilised to counter any Chinese offensive.[48] In fact, these programmes have been instrumental in creating a harmonic relationship between the border people and the armed personnel deployed along the India–China border.

It is said that a contented and happy border people is the greatest strategic asset a country can have. In this context, it is the duty of the Indian government country to make the people of Tawang happy, contented, and orient socially, culturally, and economically with India instead of Tibet or China. Keeping this objective in mind, the union government had decided to develop the border areas and launched the Border Area Development Programme (BADP) along the India–China border in 1998–99.[49] The aim of the programme is to remove the critical gaps in physical and social infrastructure needs and to strengthen the economic conditions and raise the standard of living among the inhabitants of the remote border areas. The programme has positive but limited impact on the creation of basic social and economic infrastructure in the border areas. A number of factors such as meagre financial resources, institutional inadequacies, corruption, lack of peoples' participation, and political interference have been responsible for the suboptimal performance of the programme.[50]

In view of these failures, the Government of India, in 2015, decided to redouble its efforts and develop model/smart villages with cluster approach closer to the border. The plan is to develop one such model village in a block within a time frame of 3–4 years. The objective of the scheme is to develop the smart village as the centre of socio-economic activities in due course of time which will help in preventing the out migration of population from the border areas to the hinterland. The idea is if people stay near the border, then the difficulties of informal surveillance of the border areas as well as logistics would be easy to surmount and help forces secure the border better.[51]

## Conclusion

Border management approach has been witnessing paradigmatic changes over the last couple of decade world over. Security of the international border and its management is increasingly being viewed as a shared responsibility between the union and provincial governments with active participation from the local community. In fact, it is often felt that effective border management is possible only with the active support of the border community. However, the fact that borders are at the periphery of a state and often remote, local people are, more often than not, subjected to powerful centrifugal forces. The situation is compounded further if the border areas are underdeveloped and the quality of life is poor. In such conditions, people residing in these peripheral areas develop a sense of alienation and refuse to cooperate with the state's endeavours of nation building, thus jeopardising the security of the country.

Given that Tawang is underdeveloped and lacks robust social and economic infrastructure and given that China covets the tract and is always in the lookout to exploit any internal dissensions between the local people and the Indian state, the Indian government has always been mindful of the sentiments of people of Tawang and has made all efforts to win their hearts and minds and make them active participants in the security and management of the border. The training of local communities in self and village defence, patrolling, and intelligence gathering through agencies such as the SSB, the Homes Guards, and the Lok Sahayak Sena, launching of development programmes in border areas, establishment of separate department for the promotion of language and tradition of the Monpas, etc., are some of the steps taken to integrate the local community with the country. These initiatives have also made them stakeholders in maintaining peace and security in the strategically sensitive border region. That the people of Tawang remain loyal to India and support it in the nation building project is testimony to the success of these efforts.

India, however, should not be complacent because of the success story and should make continuous efforts to strengthen the bonds between the Monpas and the rest of the state as well as the country. In this respect, firstly, a healthy relationship between the local people and the security personnel, who are present in substantial numbers in Tawang, should be fostered through sustained community interactions programmes. Security personnel should be routinely educated about the culture and language of the local people as well as sensitised about the hardships they face. At the same time, the local people should be educated about the strategic importance of their area and the constraints faced by the security personnel and the rationale of their operations. Secondly, committees comprising the security personnel, village headman, and prominent persons should be constituted for consultation and coordination regarding various issues of border management. Thirdly, the local community should be trained to act as eyes and ears for the security and intelligence agencies deployed at the border. Last but not least, the local people should be empowered and their participation should be elicited in the planning and execution of various development schemes so they become active stakeholders in a peaceful and secure border.

## Endnotes

1. 'General Chief of Staff's ' Note on the North-East Frontier', on 1 June 1912', as quoted in Noorani (2011, p. 184).
2. For details of the footpaths see, Sinha and Athale (1992).
3. 'Ex-Chinese diplomat says concession on Tawang can resolve border dispute, India says not possible', *The Indian Express*, Beijing, March 3, 2017, viewed July 16, 2018, https://indianexpress.com/article/india/arunachal-pradesh-tawang-china-india-border-dispute-4552526/.
4. Sibal (2017).
5. *Management of Indo-China Border*, Ministry of Home Affairs, viewed July 18, 2018, https://mha.gov.in/sites/default/files/INDO%20CHINA_05052017.pdf.
6. Noorani (2011, p. 186).
7. Marshall (2005).
8. Guyot-Réchard (2017, pp. 71–72).
9. Pradhan (2008).
10. *The Dalai Lama-Biography and Daily Life*, viewed July 19, 2018, https://www.dalailama.com/the-dalai-lama/biography-and-daily-life/birth-to-exile.

11. Dutta (2008).
12. 'Arunachal Pradesh is our territory: Chinese envoy', *Rediff.com*, November 14, 2006, viewed July 3, 2018, http://www.rediff.com/news/2006/nov/14china.htm.
13. 'Question no. 767: Indo-China Border Dispute', *Lok Sabha*, February 7, 2018, viewed July 18, 2019, http://www.mea.gov.in/lok-sabha.htm?dtl/29424/QUESTION+NO767+INDOCHINA+BORDER+DISPUTE.
14. Chinese incursions into India rose in 2017: Government data, *The Times of India*, New Delhi, February 5, 2018, viewed July 18, 2018, http://timesofindia.indiatimes.com/articleshow/62793362.cms?utm_source=contentofinterest&utm_medium=text&utm_campaign=cppst.
15. *Tawang*, viewed July 18, 2018, https://tawang.nic.in/about-district/.
16. Data is derived from the Official website of Tawang district, viewed on July 18, 2018, https://tawang.nic.in/.
17. *Baseline survey of Minority Concentrated Districts-Tawang*, viewed December 2, 2017, https://icssr.org/Tawang,%20Executive%20summary.pdf.
18. Ishar (2018).
19. Pillai Rajagopalan and Prakash (2013).
20. *A Synopsis of Tawang District*, viewed July 18, 2018, https://cdn.s3waas.gov.in/s39b70e8fe62e40c570a322f1b0b659098/uploads/2018/07/2018070318.pdf.
21. Ishar (2018).
22. Author's interview with local people during a field visit to Tawang.
23. "Monpas push for council—Talks with PM to preserve unique culture & religion", *The Telegraph*, April 9, 2008, at https://www.telegraphindia.com/1080409/jsp/northeast/story_9111048.jsp (Accessed on July 19, 218), also see, "MARDC Meets Rajnath to Persist Demand on Mon Autonomous Council", *Northeast Today*, August 26, 2015, at https://www.northeasttoday.in/mardc-meets-rajnath-to-persist-demand-on-mon-autonomous-council/ (Accessed on July 19, 2018).
24. 'Mon residents demand autonomous council at Tawang rally', *Business Standard*, Tawang, November 25, 2013, viewed July 19, 2018, https://www.business-standard.com/article/news-ani/mon-residents-demand-autonomous-council-at-tawang-rally-113112501014_1.html.
25. 'Arunachal to pursue Centre for autonomous district councils', *The Times of India*, Itanagar, September 24, 2013, viewed July 19, 2018, https://timesofindia.indiatimes.com/city/guwahati/Arunachal-to-pursue-Centre-for-autonomous-district-councils/articleshow/22964832.cms.
26. Panchayati Raj Department, *Growth of Panchayati Raj in Arunachal Pradesh*', viewed July 19, 2018, http://www.panchayatiraj.arunachal.gov.in/index.php?option=com_content&view=article&id=68&Itemid=114.
27. Barooah Pisharoty (2017).
28. 'Plea for inclusion of Bhoti language in 8th Schedule', *The Assam Tribune*, Itanagar, November 20, 2016, viewed July 19, 2018, http://www.assamtribune.com/scripts/detailsnew.asp?id=nov2116/at058.
29. Government of Arunachal Pradesh, *Karmic and Adhyamik (Chos-Rig) Affairs*, viewed July 19, 2018, https://karmikadhyatmik.in/.

30. 'Xi tells Tibetan herders to protect Chinese territory near Arunachal border', *The Hindustan Times*, October 29, 2017, viewed July 19, 2018, https://www.hindustantimes.com/india-news/guard-chinese-soil-xi-jinping-tells-herdsmen-from-arunachal-border/story-YdGdwV4LfpkYnGjCBcfB3M.html.
31. Ibid.
32. United Nations Development Programme, *Border Management & Border Communities in the Sahel*, viewed July 18, 2018, https://info.undp.org/docs/pdc/Documents/NER/Border%20Management%20%20Border%20Communities%20in%20the%20Sahel%20(Pro%20Doc)%20v2.pdf.
33. Ibid.
34. *Baseline survey of Minority Concentrated Districts-Tawang*, viewed December 2, 2017, https://icssr.org/Tawang,%20Executive%20summary.pdf.
35. Rose (1967).
36. Shukla (2012).
37. Ibid., pp. 418–420.
38. Nehru's five principles towards Tribals of NEFA, as quoted in n. 9, pp. 119–120.
39. Ibid. Also see Shukla and Shukla (2012).
40. 'Mon residents demand autonomous council at Tawang rally', *Business Standard*, Tawang, November 25, 2013, viewed July 19, 2018, p. 31, https://www.business-standard.com/article/news-ani/mon-residents-demand-autonomous-council-at-tawang-rally-113112501014_1.html.
41. Ibid., pp. 65–66.
42. Ibid.
43. Ibid.
44. Mullik (1971).
45. *Sashastra Seema Bal*, Ministry of Home Affairs, viewed July 18, 2018, http://www.ssb.nic.in/index1.aspx?lsid=39&cid=0&lev=2&lid=11&langid=1.
46. Ibid.
47. 'Arunachal to pursue Centre for autonomous district councils', *The Times of India*, Itanagar, September 24, 2013, viewed July 19, 2018, p. 62, https://timesofindia.indiatimes.com/city/guwahati/Arunachal-to-pursue-Centre-for-autonomous-district-councils/articleshow/22964832.cms.
48. *Baseline survey of Minority Concentrated Districts-Tawang*, viewed December 2, 2017, p. 493. https://icssr.org/Tawang,%20Executive%20summary.pdf.
49. Ministry of Home Affairs, *Border Area Development Programme (BADP)*, viewed July 20, 2018, http://mha1.nic.in/BADPHome.html.
50. PEO Report No. 229 (2015), *Evaluation Study on Border Areas Development Programme (BADP)*, Niti Aayog, New Delhi.
51. Department of Border Management, *Modified Guidelines of the Border Area Development Programme (BADP), June, 2015*, June 29, 2015, p. 16.

## References

Barooah Pisharoty, Sangeeta. 2017. When It Comes to Renaming Places in Tawang, China Is Not Alone, *The Wire*, Tawang, May 23, 2017, viewed July 19, 2018, https://thewire.in/culture/not-china-alone-renamed-places-tawang.

Dutta, Sujit. 2008. Revisiting China's Territorial Claims on Arunachal. *Strategic Analysis* 32 (4): 557.

Guyot-Réchard, Bérénice. 2017. *Shadow States: India, China and the Himalayas 1910–1962*. Cambridge: Cambridge University Press.

Ishar, Brig J.S. 2018. *History of Project Vartak*, viewed July at 18, 2018, http://bro.gov.in/WriteReadData/linkimages/8903043223-project%20history.pdf.

Marshall, Julie G. 2005. *Britain and Tibet: 1765–1947*, 462. London: RoutledgeCurzon.

Mullik, B.N. 1971. *My Years with Nehru: The Chinese Betrayal*, 491. Bombay: Allied Publishers.

Noorani, A.G. 2011. *India-China Boundary Problem 1846–1947*. New Delhi: Oxford University Press.

Pillai Rajagopalan, Rajeswari, and Rahul Prakash. 2013. Sino-Indian Border Infrastructure: An Update. *ORF Occasional Paper No. 42*, New Delhi, May 2013.

Pradhan, R.D. 2008. *Dragons's Shadow Over Arunachal: A Challenge to India's Polity*, 32–33. New Delhi: Rupa.

Rose, Leo E. 1967. *The North-Est Frontier Agency of India*, 5. Berkeley: Institute of International Study.

Shukla, Sonia. 2012. Forging New Frontiers: Integrating Tawang with India, 1951. *China Report* 48 (4): 413–415.

Shukla, Ajai, and Sonia Trikha Shukla. 2012. The 1962 War: How China Lost the Battle for Hearts, *rediff.com*, October 30, 2012, viewed July 18, 2018, http://www.rediff.com/news/slide-show/slide-show-1-the-1962-war-how-china-lost-the-battle-for-hearts/20121030.htm#1.

Sibal, Kanwal. 2017. India Need Not Worry About China's Bluster on Tawang and the Dalai Lama. *The Wire*, April 9, 2017, viewed July 16, 2018. https://thewire.in/diplomacy/122468.

Sinha, P.B., and A.A. Athale. 1992. *History of the Conflict with China, 1962*, 161–163. New Delhi: Ministry of Defence.

**Pushpita Das** is a Research Fellow and Coordinator of the Internal Security Centre at the Manohar Parrikar Institute for Defence Studies and Analyses (IDSA), New Delhi, India. Her areas of research include border security and management, coastal security, drug trafficking, migration, and India's Northeast. At IDSA, she has been researching on India's approach towards the management of its international borders. Also, she has been co-opted as an expert by the India's National Security Council Secretariat (NSCS) as well as the Ministry of Home Affairs (MHA) on projects on Coastal Security. She has written extensively on illegal migration and coastal security, etc. including two edited books titled: India's Border management: selected documents (2010), and Northeast India: New Vistas for Peace (2008).

The manufacturer's authorised representative in the EU is Springer Nature Customer Service Centre GmbH, Europaplatz 3, 69115 Heidelberg, Germany. If you have any concerns regarding our products, please contact ProductSafety@springernature.com

Printed and bound by CPI Group (UK) Ltd, Croydon, CR0 4YY
30/09/2026
02225574-0019